COMMON GROUND

COMMON
GROUND

Justin Trudeau

HarperCollins*PublishersLtd*

Common Ground
Copyright © 2014 by Justin Trudeau.
All rights reserved.

First edition

Page 335 constitutes a continuation of this copyright page.

The title of Chapter 4 is taken from the Robert Frost poem
"Stopping by Woods on a Snowy Evening," from *The Poetry of Robert Frost*,
edited by Edward Connery Lathem. Copyright 1923, © 1969 by Henry Holt and
Company, Inc., renewed 1951, by Robert Frost.

HarperCollins books may be purchased for educational, business,
or sales promotional use through our Special Markets Department.

HarperCollins Publishers Ltd
2 Bloor Street East, 20th Floor
Toronto, Ontario, Canada
M4W 1A8

www.harpercollins.ca

Library and Archives Canada Cataloguing in Publication
information is available upon request

ISBN 978-1-44343-337-2

Printed and bound in the United States of America
RRD 9 8 7 6 5 4 3 2 1

Dedicated to my best friend, partner, and soulmate.
Thank you for all you do, and for all you are.
Je t'aime, Sophie.

Contents

Prologue

———

I N THE KITCHEN AND FAMILY ROOM OF OUR HOME IN
Ottawa, there are photos wherever you look. Plastered
on the fridge, framed on shelves and countertops,
hung on the walls. There are official recorded moments
mixed with favourite family snapshots. Sophie with all the
groomsmen at our wedding; Xavier's school photo; the four
of us in Haida Gwaii on our most recent trip to B.C. before
Hadrien was born; me posing with constituents in Papineau;
my brothers, Sacha and Michel, and me riding our bikes on
the driveway of 24 Sussex; my mother, Margaret, smiling
with her grandchildren. They each spark special memories
and have meaning to us. But the group of photos that never
fails to catch my eye is a trio assembled by a good friend of

ours. These photos do more than bring back good memories; they tell a story.

The first picture shows a middle-aged man in the stern of a canoe, his paddle at the ready and a big smile on his face. The canoe is riding over a patch of rough water, and the man is watching a boy seated in the bow of the canoe who is handling his paddle with only promising skill. The man is my father and the boy, of course, is me. We are on the water on a mild spring day. The smile on my father's face suggests that he could not be more content. I suspect this is true because he is taking me on a special journey, a rite of passage that he would conduct for all of his sons.

Each of us—Sacha, Michel, and I—made this same journey over these rapids with my father. We were barely walking before Dad put a paddle in our hands and initiated us into the techniques of the voyageurs. Under his watchful eye, we'd work up to this small set of rapids that marked the outflow of Harrington Lake in the Gatineau Hills. My father didn't want us to enjoy a tranquil ride; he wanted us to face a challenge, to be involved in the journey, to help take control of things in some small way. He wanted us to have fun.

In the next photo, two men are riding an inflatable craft through water far more challenging than in the first photograph. In fact, they are in some serious whitewater rapids. The older man, sporting a somewhat scruffy beard, is in the front of the craft with his kayak paddle across his legs. He looks somewhere between exhilarated and alarmed at the boat's dangerous angle and the treacherous water around them. Behind him, in the stern, the much younger man is focused on staying clear of the large rocks nearby.

It's the same two people in both pictures, taken twenty years apart. In the second photo, my father is enjoying the ride and I'm guiding the boat, both of us totally engaged in the moment. The pictures are a touching measure of the passage of time and the effects it has on all of us.

The third picture shows—surprise!—another canoe. This one is red and shiny, it glides on glassy calm water, and I am again seated in the stern. Sophie waves to the camera from the bow. Behind her, Ella-Grace mimics her wave while Xavier watches calmly from the middle seat. The photograph captures one of our many excursions in the canoe with the children. This one, taken above Miles Canyon on the Yukon River, is significant because it would mark our last summer together as a family of four: our son Hadrien was born the following winter.

My father's presence dominates the first two pictures, and I like to think he is in the third photo as well, this time in spirit. It is well known that he loved canoeing. It took him outdoors, it challenged his sense of independence and survival, and it connected him with his roots as a young man, as a gifted athlete, and as a Canadian. He loved any opportunity to pass at least part of a day paddling across water, charging down a ski hill, or exploring a hiking trail. He was as skilled an outdoorsman as I ever expect to know.

These photos are a testament to the march of years,

but they are also the ones that resonate deeply and make me miss my father most. It was when we paddled or hiked together back then that we felt closest as a family. The city was where the stress of work and politics would sometimes beat his family down. The outdoors was where we relaxed by getting in touch with who we were and not who others wanted us to be. Together, we learned to face down obstacles and overcome our fears and we developed an endless appreciation for our country and its great natural beauty.

Today, I can no longer grab a snowboard or a paddle and a life jacket on a whim and lose myself on a mountain or a river for hours or days at a time. Sophie and I have to carve out those moments for our family, on vacation or on much-anticipated Sundays. However, the lessons of my youth remain alive in me, and they are what Sophie and I want to pass on to our children. Xavier, Ella-Grace, and Hadrien are the centre of our world and the reason we have embarked on this journey together.

I HAVE HAD THE EXTRAORDINARY OPPORTUNITY TO explore this nation at many points in my life—as a boy travelling with my father, as a young man going west for the mountains and teachers' college, as the head of Katimavik, and now as a father and as a politician. Every journey has served to remind me of the kind of country we live in, the kind of physical distances we have to bridge, and the kind of abundant gifts that come with this land. Maps can't provide any idea of the real scope of Canada, and air travel minimizes everything that our country offers. You can't appreciate the sweep of the Prairie breadbasket or the engineering achievement of Rogers Pass from thirty thousand feet. You need to be at ground level, where you can not only explore the land but meet the people who cherish the land as much as I do.

Too many Canadians emphasize their regional differences and forget the things that unite us. We are one people who speak two official languages and share a host of others. For all our differences of culture, history, and geography, we are bound together by shared values that define the Canadian identity. I have a deep-seated love and respect for Canada and recognize that we have extraordinary potential. Everything about my life has emphasized and reinforced that fact. Everything I propose to do in my political career is built on that premise.

However, it's a potential that is easily wasted and, once gone, isn't easily recovered. The last few years have seen this country's potential greatness fade in the shadow of divisive politics and a focus on seizing power for its own sake. That's not what Canada needs, nor what Canadians want. Our country was built on better goals than that, guided by a vision that was both unique and encouraging to people all over the world.

This risk to our potential is among the reasons that led me to enter politics and to make my case for a different approach to guiding Canada forward. In many ways, my approach reflects the circumstances of my upbringing and my awareness that we need to share not just the bounty of our land but the responsibility of protecting and enhancing that bounty. We need to both prize and hone our acclaimed sense of acceptance and inclusion and our respect for democratic values. We need to honour the priceless heritage of this broad and beautiful land, and its promise of a rich future for our children and grandchildren.

If I sound a bit rhapsodic, you'll have to forgive me. I tend to get that way about things I love and treasure. I wrote this book to explain why I feel this way about our country and how I learned to lead.

My vision for this country is very much shaped by my experiences and the influences upon me—Trudeau and Sinclair, father and mother, French and English, East and West. Just as every river is the sum of a hundred tributaries, so am I the product of many people and regions.

I am always a son, but today, I'm also a husband, a father, and a man passionate about his country. And if I wish to one day have the opportunity to lead Canada toward a future of justice, equality, and shared purpose, I feel I must tell you my story in my own words so that you can know better the man I am, far from the glare of politics. I'd like to share with you the sense of duty that propels me: to serve our country by fostering the common ground where every Canadian can find his or her own place within a strong and fair country.

CHAPTER ONE

Childhood at 24 Sussex

A FITTING BEGINNING TO MY STORY CAN BE FOUND more than a century ago in the town of Banff, on the thinly populated northeastern coast of Scotland known as Aberdeenshire. One day in 1911 a local schoolteacher and avid fisherman named James George Sinclair traipsed out to a nearby stream with some friends and dropped his line in the water. Almost immediately the group was set upon by a constable who declared that they were fishing illegally—the waterway was "owned," one end to the other, by the local nobleman.

Feudal land-use laws survived well into the twentieth century in Scotland and elsewhere in Europe, and the penalties for violators could be severe. If James was caught trying to pilfer the local lord's fish again, the constable warned, it would mean jail time for him.

As James and his friends packed up their gear and headed home across the meadow, he grumbled, "If I canna fish, I canna live." One of James's companions began describing a wide-open land to them, a bonnie place where the forests teemed with game and "no nobleman owns the fish." He'd read about it in a book, the fellow said. A wonderful place it was, more than four thousand miles away, across the Atlantic and on the far side of Canada. A place called "British Columbia."

A few months later, James George Sinclair, his wife, Betsy, and their three-year-old son, Jimmy, were aboard a boat sailing to Canada. They found much more than fish in British Columbia. Their new home was a land of opportunity where hard work paid off, whatever your accent or ancestry. Over the next half century their son Jimmy grew up to earn a degree in engineering, become a Rhodes Scholar, serve as an RCAF officer in World War Two, be elected an MP, serve as a cabinet minister, have a successful business career—and remain all his life, like his father before him, an avid fisherman.

He and his wife, Kathleen, named the fourth of their five daughters Margaret. Today she lives in Montreal; she's my mom.

IN SEPTEMBER 1941, WHILE JIMMY SINCLAIR HAD THE particular distinction of serving his first term as MP for the riding of Vancouver North while commanding an RCAF squadron in North Africa, a French-Canadian intellectual embarked on an extraordinary sixteen-hundred-kilometre canoe expedition from Montreal to James Bay, retracing the seventeenth-century journey made by the coureurs de bois who founded the Hudson's Bay Company. The trip attracted some media attention; under the headline "Students Went on a Pleasant Voyage," a local newspaper listed the six canoeists, including one by the name of Pierre E. Trudeau.

It was an arduous journey. For my father, that was precisely the point. "I shot the rapids while the others portaged," he wrote in a letter to a friend. "The food began to give out, the portages were impossible, the rapids dangerous . . . In a word, life was becoming beautiful." This was the lens through which my father saw his native Quebec—as a proud and magnificent place full of rugged beauty. He always believed that the province's defining spirit emerged as much from the land as from the language and culture.

As a family, we've always had a strong connection to the water. In fact, water plays a role in my very first memory. I was not quite two years old, bundled up in a snowsuit and sledding with my father at Harrington Lake, the government-owned prime ministerial residence in Gatineau Park, which was one of my parents' favourite places to spend time together. It was December 1973, and the lake was not quite frozen over. My mother stood at the top of a hill, ready to burst with the imminent birth of my brother Sacha, and cheered us on as my father went up and down the slope with me on a sled. Each swift descent ended near the stream that flowed out of the lake, the one I would later paddle down.

After a few turns, my father satisfied himself that the run was safe and decided I should have a go by myself. From the top of the hill he gave the sled a push, and off I went down the slope while he and my mother looked on. Almost immediately, my dad saw a huge problem. When he and I were aboard the sled together, our combined weight was enough for the sled's runners to break through the icy crust and slow us down. But with just me on board, the sled skimmed lightly on the crust more like a skate and began gaining speed, heading directly for the stream. As my father bounded down the slope in hot pursuit of me, my mother stood atop the hill terrified, shouting, "My baby, my baby!"

As young as I was, I clearly recall the ride ending with the sled half-buried in the sandy shore and my outstretched hands wrist-deep in the ice-cold water. I was wearing blue knit mittens, and my principal concern was that I had gotten them soaked. "Fall down river, mittens wet!" I cried out to my father, half-delighted and half-surprised, when he arrived to rescue me. He scooped me up with one hand, grabbed the sled with the other, and carried me back up the hill. It was a significant day: I had been baptized an outdoorsman.

Before this adventure, however, was the eventful time of my birth. Sir John A. Macdonald was the last prime minister to have a child in office. My father and mother both embraced the goals of the new feminist movement that was revolutionizing the way men and women approached their roles as parents. However, they were born three decades apart, and the difference in their ages was something that was not easily overcome. To put that in perspective, my father was born in 1919, the year that women gained the right to stand for federal office in Canada.

In 1971, the Ottawa Civic Hospital still excluded husbands from accompanying their wives in the delivery room. My mother was furious when she heard about this. If her husband couldn't be at her side in the hospital when she gave birth, she would have the baby—that was me—at 24 Sussex. When word of my mother's protest reached the hospital's board of directors, they promptly abolished the old-fashioned restriction, followed by other hospitals in Ottawa and eventually across the country. On Christmas Day, my father was at my mother's side when I came into the world. It was, I am told by reliable sources, an easy and uncomplicated delivery. And I like to think that, along with my father, I helped my mother strike a blow against old-school patriarchal thinking.

My brother Sacha arrived two years after me, and Michel followed less than two years later, so we were close in many ways. We were constant playmates—chasing, teasing, getting into scrapes. Actually, we were rough-and-tumble little

lion cubs. I taught Sacha to wrestle when he was still in dia-
pers, and Sacha was rolling around with Michel when he was
still a toddler. Taking a cue from all that energy, my parents
put tumbling mats in the basement of 24 Sussex, eager to
see us burn off our boyish hyperactivity in a wholesome way.

Harrington Lake in those days was like the setting for
a Hardy Boys novel, a place that begged for adventure. My
father, to our delight, always seemed to encourage the idea.
An old farmstead with an abandoned barn could be explored
nearby. Halfway down the lake, past an old mica mine, sat an
unused boathouse where my brothers and I would sun ourselves in the summer. About a hundred metres offshore was
a tiny island that was the locus of our own rite of passage.
When each of us turned seven years old, we determined we
would swim out to the island and back again.

It was an example of our father's encouragement to
always test our physical boundaries that he agreed to this.
Of course, he guided and protected us—when we attempted
this ritual he was there, swimming alongside us, to the island
and back.

He also liked to surprise us. He would pull out topo-
graphical maps of Gatineau Park, place his finger on a spot
and say, "*On va là.*" A half-hour later, we would all find ourselves scrambling to keep up with him and our mother as
he marched confidently into the wilderness. His sense of
direction was excellent, and we never got lost. But the same
wasn't true for other visitors to the area. Occasionally some
confused hiker would stumble upon us and find himself get-

ting directions from the prime minister of Canada. When I look back on such episodes now, they do seem surreal. But as a young child, the prime minister assisting hikers lost in the Gatineau Hills seemed perfectly normal.

A change in season didn't stop our outdoor explorations and family excursions. Snow on the ground meant many things. We all began skiing at a young age, but at Harrington Lake, we'd usually strap on snowshoes and head out the door. These weren't the modern lightweight designs available today. We wore the old wooden teardrop-shaped variety, which looked a bit like tennis racquets and were strung with catgut (which, our father assured us, doesn't really come from cats). While trekking through the wilderness, my father, always in French, would spin tales of Albert Johnson, the Mad Trapper of Rat River, an infamous Depression-era criminal who led the RCMP on a nearly 250-kilometre manhunt through the Northwest Territories and into the frozen Yukon wilderness. This, naturally, inspired us to take turns playing the Mad Trapper, heading into the Gatineau countryside to see if we could evade capture by other family members

Tracking someone in snowshoes is easy if he walks in a straight line. The idea was to confuse the pursuers by walking in circles, branching off and doubling back, following a figure-eight pattern or even swinging from a tree branch to create a break in the trail. We loved this game, and it kept us going for hours.

After leading an RCMP posse on a hunt for more than a month, the Mad Trapper was shot to death by the Mounties

on a frozen bend of the Eagle River. Our pursuits, by contrast, usually ended with my father breaking up and sharing
a bar of dark chocolate.

I WAS EIGHT OR NINE YEARS OLD BEFORE I HAD A FIRM
grasp of my father's career and what he did when he wasn't
at home with us. My mother loves to tell the story of how I
referred once to my dad as "the boss of Canada." But what
did that mean exactly? My friends' parents performed work
I could understand—they worked in stores, or looked after
people as doctors, or talked on the radio. I could wrap my
head around that kind of work. The concept of public service was much more abstract, more difficult to understand.

The subject came up one day when I asked my father
something about our house and he replied that we didn't
own it the same way we owned our clothing or books. We
didn't? That was strange. We *lived* at 24 Sussex, so why
wasn't the house ours? His explanation was that it belonged
to the government, which just confused me more. Wasn't
my father *in charge* of the government? Didn't that make it
all his? Then, in 1979, the Liberals lost the federal election.
Almost overnight 24 Sussex was no longer our home, and we
packed up and moved a few blocks away to Stornoway, the
official opposition leader's residence. That's when I understood that the real boss of Canada was the Canadian people.

With time I began to grasp some of the more complex
issues my father dealt with, and he made a point of draw-

ing my attention to major events and their importance. For understandable reasons, he spoke to his young sons about the establishment of the Charter of Rights and Freedoms in 1982. I was ten years old at the time, old enough to be familiar with the basic principles of democracy, including the notion that governments rise and fall according to the voters' will. In explaining the importance of the Charter, my father, who had envisioned such a document from the days when he had served as minister of justice in the Pearson government of the 1960s, pointed out that some rules were too important for the government to override.

The idea that a majority of the people—or, given our electoral system, sometimes far less than a majority—could use the government's immense power to restrict minority rights appalled my father. He called this "the tyranny of the majority." The way he explained it to us as children was to say that, for example, right-handed people, who make up a large majority of the population, shouldn't be allowed to make laws that hurt left-handed people just because they are a minority.

Dad was a member both of a linguistic minority and of a generation that had seen people harness and marshal the state's power to do unspeakable things to each other the world over. He had fought his whole life to build and shape, in Canada, a country of unparalleled diversity: of religion, ethnic origin, and belief. For diversity to work, people have to be free. The Charter was his way of ensuring that it would be impossible for any group of Canadians to use the government to unduly restrict basic freedoms for any other group of Canadians. His core value was classically liberal in this sense. It is a value I share, and believe in equally deeply.

In the ensuing years, the Charter of Rights and Freedoms became the vehicle for an unprecedented expansion of individual freedom in Canada. It has been used to strike down arbitrary laws that restricted Canadians' choices in the most private and intimate aspects of our lives. Thanks to the Charter, Canadians are no longer discriminated against in their workplaces because of their sexual orientation, nor are they prevented from marrying the person they love just because they each happen to be of the same sex. Because of the Charter, women have gained the right to control their reproductive health. Other aspects of the Constitution Act were aimed at the same end. First Nations, for example, have used section 35 to establish in law rights that have been infringed by governments since European contact.

Since entering Parliament in 2008, I've often thought about what the Harper years would be like without the Charter of Rights and Freedoms. Mr. Harper and his party are not fans of the Charter. They refused to celebrate its thirtieth birthday. They rarely mention it, and the Supreme Court has used it to curb many of their most autocratic tendencies. Personally, I believe it comes down to a key difference between liberal and conservative ideas of freedom. The liberal idea is that all individuals, regardless of their background or belief, hold the same basic rights and freedoms, and that the Constitution ought to protect them from the powerful forces that would restrict—and in extreme instances remove—those rights. The conservative idea seems to me to be much more focused on giving people and groups who have power the freedom to use their power however they choose.

I believe very deeply in the liberal idea of freedom. In the spring of 2014, I would announce a firm stance in favour of a woman's right to choose. It was a big change for some of my parliamentary colleagues. Previously, the Liberal Party considered this right to be subservient to the freedom of an individual MP to vote in Parliament according to his or her religious beliefs. As someone who was raised Roman Catholic, and who attended a Jesuit school, I understand that it is difficult for people of deep faith to set their beliefs aside in order to serve Canadians who may not share those beliefs. But for me, this is what liberalism is all about. It is the idea that private belief, while it ought to be valued and respected, is fundamentally different from public duty. My idea of freedom is that we should protect the rights of people to believe what their conscience dictates, but fight equally hard to protect people from having the beliefs of others imposed upon them. That is the difference between the views expressed by a citizen and the votes counted in Parliament. When MPs vote in Parliament, they are not just expressing an opinion; they are expressing a will to have all other Canadians bound by their opinion, under law. That is where we need to draw a firm line. I am confident that my father, were he around today, would agree.

His job may have been unique, but my dad, whom we called Papa because we always spoke with him in French, was in many ways like most fathers. He would joke with us, play games with us, and, as a special treat, sometimes take us along to work. This usually meant several hours of Sacha, Michel, and me playing tag or hide-and-seek on the third floor of the Centre Block of the Parliament Buildings. To this day, I can't pass certain rooms or stairwells in the building without the memory of those times flooding back to me.

I got my closest look at the day-to-day work my father performed as prime minister not in Ottawa, where he maintained a firm barrier between his roles as both prime minister and parent, but when we travelled around the country or abroad. In Ottawa, aside from our appearance at ceremonial occasions such as Remembrance Day and Canada Day, we had little exposure to his public duties. But when we accompanied him beyond Ottawa, things were different.

When it was my turn to travel overseas with him, I would often sit munching on a breakfast muffin in some hotel while my father received detailed briefings on the day's meetings from people like Bob Fowler, his foreign policy advisor, and Ted Johnson, his executive assistant. I would sometimes attend evening events as well. This gave me the opportunity to meet international leaders such as British prime minister Margaret Thatcher, German chancellor Helmut Schmidt, and Swedish prime minister Olof Palme, who gave me a reindeer-hilted hunting knife that I treasure to this day.

Sometimes I had a front-row seat at events of major

importance, such as the time I was with him on a tour of Canadian military bases in western Europe in 1982 and a bulletin announced that Soviet leader Leonid Brezhnev had died. The next day we were on our way to Moscow for the funeral.

We were met at the airport by Canada's ambassador to the Soviet Union, Geoffrey Pearson, who briefed my father during the drive to the hotel. Much of the discussion, I remember, centred around the question of who would succeed Brezhnev. Passing through Moscow, I watched evening fall over the dark, sullen city while my father carried on a long and detailed discussion of internal Soviet politics in which he evenly matched a diplomat who was stationed in Moscow. It was yet another confirmation for the boy I was that my father pretty much knew everything.

There's a limit to how much a child can process when it comes to arms control or trade agreements. But one thing I learned to appreciate was the concept that in foreign relations, relationships are vitally important. I was struck by how my father's briefings were often as much about the personalities of his counterparts as about the issues.

This became especially interesting when I was able to watch leaders from other countries meet with my father. Sometimes they seemed so different that I marvelled that they could interact productively with one another. Like Ronald Reagan.

I was nine when the president arrived for lunch with my father at 24 Sussex. It was pretty clear that day that something momentous was happening, because RCMP officers were positioned at ten-foot intervals all around the property, which was more security than I had seen on the lawn before or since.

When the charismatic U.S. president entered, my father introduced me and suggested that the three of us relax in the sun room before the two leaders had lunch. Reagan smiled warmly at me as we sat down and asked if I'd like to hear a poem, which made my father cock his head with interest. He loved poetry and often assigned us verses to memorize from works such as Racine's *Phèdre* and Shakespeare's *The Tempest*. But Reagan had different tastes. Instead of classical verse he launched into Robert Service's "The Shooting of Dan McGrew" ("A bunch of the boys were whooping it up in the Malamute saloon . . .").

I was enchanted by the verse. My father was somewhat bemused both by the mildly inappropriate subject matter for a nine-year-old and by the predictably appropriate choice of verse by the cowboy/actor president. Still, it made an impact; I was impressed enough to memorize it and other narrative poems that my father would never have taught me, from "The Cremation of Sam McGee" to Alfred Noyes's "The Highwayman."

Equally memorable were the times aboard the government Boeing 707 used for international trips. The front section of the aircraft had eight large seats, facing each other in groups of four. Behind them were two long benches, where my father and I slept during long flights. A wall separated this section from the rest of the plane, which was for staff, security, and press. I would sometimes go back to speak to the people I knew, because my father often worked on planes and there were no brothers to play with. But as interesting as the conversations were, I kept my visits to this section brief. Smoking was still allowed on planes then, including government planes, and the opaque fog of smoke that enveloped that area of the aircraft left me coughing.

The most valuable part of these trips with my father was the chance to watch how he made decisions. He was always asking questions and challenging the people around him about their opinions. He would rarely discuss his own views in any detail until everyone else had had their say, which was in contrast with his public image as an almost autocratic decision maker. Any decision made by my father was the result of a process that had involved many voices, and which sometimes had taken weeks or months. The decision-making model I learned during those 707 flights has come to inform my own leadership style.

All of this was the context in which I grew up. What stands in the forefront of my mind, however, was how the five of us lived as a family while in Ottawa, and how devoted my parents were to us.

DESPITE THE DEMANDS ON HIS TIME, DAD WAS AN engaged, hands-on father who took great joy in his children. He found satisfaction in performing parental chores, tending to us in the night when we were infants or repairing our bicycles and assembling Christmas toys when we got older. He spun bedtime stories, *en français, bien sûr*, about Jason and the Golden Fleece or Paris and Helen of Troy, or scared our pants off with the story of Polyphemus and his cave. In the daylight hours, he introduced us to almost every physical activity available, though team sports like soccer, football, and hockey didn't appear to interest him. He taught us to sail, to rock climb, to use a gun and archery equipment, to navigate the outdoors, to swim and dive and rappel and, of course, to ski. At Harrington Lake we spent at least four hours each day in some kind of outdoor activity whether it was hot or cold, dry or drenching. My dad had a great saying: "There is no such thing as bad weather, only bad clothes."

He and my mother excelled at skiing. My mother has always been a beautiful skier. As for my father, even on the most difficult hills, he outshone other skiers with his graceful style and aggressive attitude, and until he was well into his seventies he kept up with my brothers and me on the most challenging runs.

Off the skis and out of the canoe, he practised his ballroom dancing and lost himself in classical music and serious literature, sharing these enthusiasms with us as well. We were encouraged, expected even, to know history, Catholic theology, and the basis of philosophy as well as we knew how to make a parallel turn on skis and how to portage a canoe through heavy brush.

The three of us enrolled in judo classes, which helped us learn to fall and tumble, and when I was four or five, my father taught me to box, which later became something I actively pursued.

My mother insisted on broadening our horizons in other directions. In my case, she achieved only varying success. When I was just six years old, she enrolled me in ballet class. I'm a great believer in eclectic interests where culture is concerned, but being one of two boys among sixteen young girls was more than my young ego could take. My mother and the ballet teacher made a concession to my self-consciousness by allowing me to wear pants instead of tights, but it wasn't enough. I hated the whole idea and rebelled at being dragged to ballet class until the day my mother was forced to literally pull me through the door of 24 Sussex while I kicked and screamed. I seized the door frame and clung to it desperately, refusing to give in to Mom's pleas until a workman painting a railing near the door, who had watched us for a moment, said, "Come on, lady. Give the kid a break."

That did the trick. I went to ballet that day, but it was my last visit.

While my mother and father worked very well together as parents, it's known that they faced many challenges as a couple. My mother's theory is that she and my father weren't capable of having a "normal" or productive argument. There was no middle ground, so instead of any sort of gradual meeting of the minds, the dams would burst and both my parents would let loose. Over time, the unpleasant episodes between them multiplied, until their marriage fell apart.

My mother readily acknowledges that my father was an exemplary parent by always making time to spend with his children. In fact, his attitude toward parenting was decades ahead of its time. He almost always found something new that piqued our interest, some fascinating discovery to explore, or just some way to make us laugh and be happy.

Sometimes his active parenting practices came as a shock to his stodgier colleagues. When I was still a baby, Dad often would come home during the day to help care for me, racing upstairs to the nursery before he'd even taken off his coat. To make the arrangement work, he would invite his cabinet ministers to 24 Sussex for a working lunch. On one memorable occasion, he plunked me down in a baby seat in the middle of the table in the dining-room alcove to the amazement of his assembled colleagues. John Turner, my father's newly appointed finance minister, observed me for a moment and then said, "Don't worry, Pierre. Kids get a lot more fun and interesting once they get a little older." Years later, when my dad would tell this story, he still found John's comment bewildering: for him there was nothing more interesting than watching even a baby discover the world. He revelled in our first words and first steps every bit as much as in our first backflips off a diving board or on a trampoline. From my first memories of my father to my last, his love for us was clear. That fact, more than any other, is the anchor of my childhood.

And let me be honest: lots of things about being the prime minister's son were just plain fun. Like the special code names that the RCMP gave our family: my dad and mum were Maples 1 and 2, my brothers 4 and 5. I was Maple 3. All the major locations in our life had code names as well. My school, Rockcliffe Park Public School, was known as Section 81, and Section 76 was my buddy Jeff's house. Sometimes the RCMP officers would let my brothers and me take the microphone to exchange coded messages with Mounties in other cars. I remember the pride I felt the day I cracked their so-called secret code: "Alpha Bravo Charlie! You just take the first letter of each word!"

Birthday parties at 24 Sussex were especially enjoyable, a time when we would change the rambling old mansion into a playhouse for a day. Since Sacha and I shared a Christmas birthday, in mid-December we would each invite our entire class over. About forty kids would arrive, Dad would retreat to his office, and we would be free to play avalanche tag, a form of hide-and-seek in which every newly found player joins the search party until, at the end of the game, a whole pack of kids are searching for the one remaining hidden kid.

This was the side of my life that my school friends saw and sometimes envied. Occasionally, unexpected things would happen that made jaws drop. I remember a June day when I was eleven years old and playing on the driveway of 24 Sussex with my friend Jeff Gillin. A car pulled up, the door opened, and an elegant young woman stepped out carrying a gym bag: it was Diana, Princess of Wales. She and Prince Charles were touring Canada at the time and I had been told she was discreetly coming over to swim some laps in the pool at the back of the property, and so I decided it would be appropriate to greet her properly.

Jeff and I vaguely sensed that some sort of protocol should be observed, but standing there in our grubby T-shirts and jeans, we hadn't the faintest idea of what to do. Bow deeply? Salute? Instead, we dropped our bikes and sort of stood there at attention, a child's version of an honour guard as the princess passed. For me, it was an awkward moment, made worse by the fact that she was obviously a little out of sorts that we had intruded on what was supposed to be her totally secret and private time. And so as soon as she'd whisked past us (with just a hint of an eye roll), I turned to apologize to Jeff for what had happened. Jeff, his eyes as large as saucers, exclaimed, "Oh my God! That was *incredible*!"

Another incident involving Jeff occurred around the same time. He and I and a few friends were riding our bicycles through the neighbourhood, and as usual, an officer in an RCMP car trailed behind us at a safe distance. I thought nothing of it, but when one of my friends decided it would be fun to shake our security shadow, we took a sudden sharp turn through a park, went down some back streets, and looped through a meandering route back to Jeff's house—where, of course, the RCMP officer, who had guessed what we were doing, was waiting. When my friends and I finished playing, the officer escorted me home and, as he had to, submitted a report of the "incident."

My friends and I thought our efforts to lose the officer were fun. My father thought differently. In a word, he was furious. "Do you think these guys *like* having to follow around an eleven-year-old kid?" he demanded. "Their job is to keep you safe so I can do my job. And here you are deliberately trying to make their job difficult . . . for *fun?*" Then he added, in that stern tone I knew all too well: "This was a total lack of respect for them. I raised you better than that."

Disappointing my father was just about the worst thing I could do as a child. I yearned, as most kids do, for his attention and approval. While he gave both often, his disapproval was a wrenching experience for me.

There were times, however, when we slipped across the line in brattiness. I don't know if Sacha, Michel, and I were less or more "bratty" than other rambunctious boys our age. I only know that both our parents, and especially our father, had zero tolerance for anything other than respectful behaviour. We may have lived in a privileged environment, but when it came to expectations and discipline we were not pampered. On the contrary.

My mother always emphasized the importance of good manners. A breach of protocol or etiquette resulted in a stern rebuke from her. "Good manners will open doors for you," she lectured, "and once a door is open, you can demonstrate your good character." She also insisted that our attitude toward and interest in other people be genuine. "Don't be phony," she said. "People can always tell if you're being phony, and when they do, they'll never fully trust you again."

The importance of being both honest and respectful of others was a pillar of the teachings my brothers and I received from both parents. When I was eight years old, my father took me to Parliament Hill, where we had lunch in the restaurant there. Looking up from my meal, I spotted Joe Clark, the leader of the Progressive Conservative opposition. Thinking to please my father, I repeated a silly joke about Joe that I had heard in the schoolyard. It failed to amuse him. In fact, it appalled him, and I received a stern lecture about how it was fair to attack an opponent's position, but I was never to make a personal attack on the individual. To drive his point home he marched me over to Mr. Clark's table, where he was sitting with his daughter, Catherine, and introduced us.

I often have wondered how my father would react to the wider use of personal attacks by some on the current political scene in lieu of any serious discussion of issues. I have no doubt he would be disgusted and, yes, disappointed in us all, and that he would find a way to express his opinions with the weight of a falling ton of bricks, but without the need to resort to the same personal attacks he was decrying.

This emphasis on respecting others, whatever their position or title, was among the most important lessons drilled into my brothers and me as children. Sometimes our parents took the initiative in drawing our attention to another's qualities and the high regard they deserved. Our house-keeper Hildegarde West, known simply as Hilda to us, was one of those people. It's difficult to describe precisely what it was about Hilda that generated so much affection toward her in the family, except to say that she radiated warmth in every direction.

One day, sparked perhaps by a comment from me or a gracious gesture by Hilda, my mother took me aside and said, "Justin, during your life you are going to meet kings, queens, presidents, all kinds of people with power and prestige. But whatever their titles, many of them will never have the worth, as human beings, that Hilda has."

Dad, if anything, was even more strict about the need for respect when dealing with others. Once, when I referred to an RCMP agent assigned to us as Baldy, the agent took it in good humour. My father, however, overhearing my comment, insisted that I formally apologize to the man then and there. The agent might have been amused by a young kid using that term in a casual manner. Dad wasn't. And he ensured that I knew and remembered it.

BEYOND THE WORLD OF 24 SUSSEX WAS THE SPIRITED western branch of the family. Visiting the Sinclairs in British Columbia was always a good escape from Ottawa and its restrictions. This is the half of my family tree that too many people overlook. Everyone knows me as the son of a former prime minister, but many forget that I am also the grandson of another extraordinary politician, Jimmy Sinclair, who, as I mentioned, had been born in Scotland and arrived in B.C. as a toddler. He provided me with one of my two middle names plus a lot of wonderful memories.

After serving as an RCAF squadron leader in Sicily, Malta, and North Africa during the war, Jimmy became a fixture in Parliament, representing the ridings of Vancouver North and Coast-Capilano and serving as fisheries minister under Prime Minister Louis St. Laurent. When his political career ended, he became president and chairman of Lafarge Cement North America, capping a remarkable career.

Jimmy was very much "a man's man," with all the charisma and outsized personality of a true old-school retail politician. As we know, my father could handle crowds and people very well, but it wasn't a natural thing for him; he had to train himself to overcome his innate shyness. For Grampa, it was all about people. Election campaigns for James Sinclair were a cottage-industry family operation, with everyone including the children pitching in to ensure success at the polls. Heather, my mother's eldest sister, remembers answering the home telephone line as a six-year-old singing, "Two-four-six-eight, who do we appreciate? Jimmy Sinclair!" Many years later, when I was running as the Liberal candidate in the Montreal riding of Papineau, it was Jimmy's door-to-door campaign style, not my dad's, that I took as my model because, among other things, it suited my personality.

Jimmy exhibited a special fondness for me. I had lots of opportunities to spend time with him, because my parents would rely on the Sinclairs to take care of my brothers and me whenever they travelled overseas for extended periods. Behind his home on Rockridge Road, in West Vancouver, Jimmy would take us through his amazing garden, which dropped behind the house all the way down the forested embankment to Cypress Creek. He transformed the embankment into an extension of the garden, creating landmarks named after my brothers and me. Here was Justin's Path, over there was Sacha's Rock, and farther along was Michel's Lookout. We would spend entire days in those magical woods with him, helping him garden, playing hide-and-seek, exploring up and down the creek.

A railroad track bordered the property, and trains ran past regularly, including the historic steam-powered Royal Hudson, which carried tourists across West Vancouver, up to Squamish, and back. Whenever it passed we would wave it out, sometimes displaying a big sign that read "O Canada," and when the engineer saw it he would blow the train's special whistle that played the first four notes of our national anthem. Jimmy had an immigrant's love of this country. And he instilled it in those around him every day.

Looking back at those scenes from Rockridge Road, we were like poster children for 1970s-era Canadian patriotism. That's why that part of the country has had a special significance to me, and drew me to it when I was in my mid-twenties, prepared to begin my career as an educator.

Jimmy would play cards with us, and we often played a game he called Bank. During the card games he would tell us that this or that hand would determine who would be the "Champ of the Western desert," which I always thought of as some empty but neat-sounding phrase. Only when I was an adult and learned about Jimmy's military past did I realize that the "desert" was the Western Sahara, where he had served his country during some of the fiercest battles of the Second World War. It gave me a chill to realize that the offhand phrase he used during a family card game was from a real theatre of war. I often think of him when I meet with veterans across the country in my current job, and I grow touched by their devotion to service and duty and by all the untold stories that lie hidden.

When Grampa Jimmy passed away in 1984, it was the first death my brothers and I experienced as a real personal loss. Hearing the news at 24 Sussex, we blubbered so loudly that one of the staff, a woman from France, finding it some-what unseemly, asked us to pull ourselves together. Needless to say, we chose not to.

On one trip west, my brothers and I journeyed to the Sunshine Coast, where we visited the home of our moth-er's elderly grandmother, whom we called Gee. She had emigrated from Britain so many years before and lived a long and quiet life in Gibsons. It is a picturesque place, and Gee whiled away her years there with books from the local library. When my mother told me stories of spending so many happy times in her childhood on that same beach, I first became aware of the passage of time. Parents age.

I had also met my paternal grandmother, Grace Elliott,

although I was far too young to recall it. Dementia had taken its toll by the time I was born, but when my mother placed me in her lap, Grace seemed to attain a moment of lucidity. "Pierre's son?" she said with tears streaming down her face. "Pierre gave us a son?!" She died about a year later.

I developed a strong attachment to the entire Sinclair family. Mom's mom, Kathleen, was an amazing woman, and I'm glad Xavier and Ella got to know her a bit before she passed away a few years ago. My aunt Janet was a strong labour activist in the airline industry, and even though she retired recently, the Vancouver airport is still the only place I get special treatment when I travel, because there I'm recognized as Jan's nephew (that's right, never any perks at PET in Montreal). My aunt Lin moved down to the States when I was little, and whenever she comes to visit with her wonderful husband, Fred, we have hilarious political conversations because Fred is really, really Republican. Betsy, the youngest, is a semi-retired RN who also runs the Brentwood Bay Nurseries on Vancouver Island with my British expat uncle, Robin. My uncle Tom, who played for the BC Lions in his youth, is also my godfather, and is married to Heather, the eldest of the Sinclair daughters, who worked as a teacher and was my mentor when I travelled west to study education. From the time she served as Jimmy's youngest campaign volunteer in the 1940s, Heather never really lost the political bug, and she still works actively in Liberal politics in Toronto, including during my leadership race. So in 2013, when I attended my first Question Period as leader of the Liberal Party in the House of Commons, she was there waving to me from the public gallery. I saluted her and directed her attention to my chest. As she squinted, I saw a smile of recognition: to honour the day, I had worn my Sinclair tartan tie.

It seemed the fitting thing to do. While my relationship with my mother over the years had its ups, downs, and then ups again, the larger connections between Trudeau and Sinclair have remained gratifyingly solid.

NOT ALL THE MEMORIES OF MY TIME AS THE SON OF A prime minister are happy. There were sad times as well, most of them connected with the difficulties in my parents' marriage.

Much has been written about their marriage and the way it ended. A lot of it is lurid and inaccurate. It's also, as you can appreciate, intensely personal to me, and I hesitated before addressing it here. In the end, I decided that if I wanted to write a book explaining how I came to be the person I am, I had no choice. Both of my parents exerted wonderful influences on me, and much of who I am today can be directly attributed to the guidance and example they provided. But like every child of divorced parents, I was shaped by their breakup as well.

In recent years, I developed a clearer understanding of the stresses that existed in my parents' marriage. One was the difference in their ages that I mentioned earlier, something that is easy to identify and blame for the problems that occurred. It's important to also remember, however, that they were two people who were very much in love with each other at the beginning of their marriage and, to a large degree, through the rest of their lives.

The element in the matter that is too rarely discussed, even after my mother's openness about it, is her lifelong struggle with bipolarity. Living your life in the public eye is a greater burden than most people can imagine. Its effect is neither insurmountable nor necessarily traumatic, but it demands that you maintain a state of mind that enables you to handle the steady pressure and periodic hassles. My father could usually revel in the hardships, taking them as a personal test or challenge to be surpassed with focus and discipline. For my mother, the experience was quite different, made difficult, even intolerable, by her condition.

Bipolar disorder is not exceptionally rare. Studies suggest that about three percent of the world's population suffers from it, equally shared by men and women and cutting across ethnic, racial, and social identities. Have a hundred Facebook friends? Chances are three of them will exhibit symptoms of bipolarity. Many mental conditions fail to receive the recognition and treatment they need and deserve. That's unfortunate. Break an arm, develop a rash, or suffer a chronic cough and you'll probably deal with it directly by seeking professional help and garnering a lot of sympathy from others. It's not the same with mental health issues. Unfortunately, even in our relatively enlightened age, illnesses associated with a mental state are not addressed as openly as they should be. Sufferers assume they will "get over it" (advice often dispensed by friends and relatives), or fear that some unexplained stigma will be attached to their condition.

My mother always had a brilliant mind waiting to shine through, and when she finally came to terms with her illness, she became an activist in the field of dealing openly with mental illness. She has spoken about it and about her personal experiences in confronting it on many occasions, sometimes with me by her side at the podium. In 2010 she wrote an exceptional memoir, *Changing My Mind*, that reflected her hard-won state of self-awareness about her illness.

One of the messages that comes through in that book is the need for people to discuss mental health issues in a candid and constructive way. It's an enlightened attitude that unfortunately was unheard of in the 1970s, when my mother began to wrestle with her bipolar condition. Had it been prevalent at the time, her years as a young mother and wife would surely not have been so agonizing.

But there was still that matter of the thirty years' difference in ages for them to contend with. Even in the absence of my mother's underlying medical issues, it would have remained a difficult barrier to overcome. My mother may have been smitten by her first glimpse of my father on the beaches of Tahiti in 1967, and he may have been equally entranced by her charm and beauty when they met again a few years later, but reality always has a way of elbowing its way into our lives. The truth was that my mother considered Pierre something of a "fuddy duddy" at times. He had become almost an icon of social liberalism by the time he married my mother, but throughout that marriage he could not escape the traditionalist mindset that had been drummed into him as a child.

My mother, on the other hand, was ahead of the social curve. The most common portrayal of her was as a "flower child," breaking free of the kind of restrictions that her husband had considered customary. Her sense of confinement, of being a well-tended bird in the gilded cage of 24 Sussex, was something she couldn't bear. "Pierre was widely seen by the rest of the world as a man who did pirouettes," Margaret wrote in *Changing My Mind*. "But what he really did was work—hour after hour. Unless it was an official occasion, we never went to the ballet or theatre. For him, this life was perfect . . . For me, it wasn't enough; I wanted, I needed, to play."

There were other complications, including religious differences. My father was a devout Catholic while my mother, though raised Anglican, has very little religious identity aside from some 1960s era flirtation with Buddhism. She could never understand the pervasive attitude of guilt that seemed to hang over my father whenever he felt he had lapsed in some way, and she was offended by the degree of intrusion that the church practised. "Even your *thoughts* can become the subject of sin and confession," she commented to me at one point. This invasion of the private sphere of one's mind in search of "thought crimes," she said, borrowing a term from Orwell, was especially disturbing to her.

From my perspective today, the commonly held story of my parents' marital breakdown is nothing but a caricature, because my father was not just the tradition-bound diehard he appeared and my mother was not entirely the totally free spirit that her actions suggest. Things are never that simple, especially with a couple as complex as my parents, and I remain amused by and exasperated with those who view their relationship—all the passion, triumph, achievements, and tragedy—in black and white, seeing it merely as a flawed union between a cool and aloof man and an exuberant and uninhibited younger woman. It was that, but also much more.

My mother frequently referred to Dad as Mr. Spock. Whenever they had an argument, my father's utterly rational approach, she claimed, was overly Vulcan. When she grew emotional and fervent, Dad would respond with logic and rhetoric in a manner that struck her as patronizing and bloodless. He, of course, would consider her behaviour exasperating when it was actually a plea for help.

My mother saw Pierre as a workaholic, a man whose identity appeared defined by his devotion to his country. In a manner, of course, it was. But his devotion to his kids was equally strong. For her part, my mother missed the highly social environment she had thrived in as one of five daughters in a vibrant and gregarious West Coast family. Throughout her childhood, the Sinclairs' West Vancouver home had been a social hub, with friends and family dropping in for impromptu drinks and dinners followed by hours of laughter, storytelling, and sharing recipes. She grew up among people whose personal goal appeared to be harvesting as much joy from life as possible, seizing the day in the manner that marks so much of life in the West.

Such a lifestyle proved impossible at 24 Sussex, a huge and drafty house that Mom called "the crown jewel of the federal penitentiary system." At other times, she compared the residence to a convent and herself to its mother superior, presiding over seven often lonely female staff who carried out cleaning and cooking duties, plus a succession of wonderful nannies assigned to help raise my brothers and me. (Among them, Diane Lavergne, Leslie Kimberley, Monica Mallon, and Leslie's sister Vicki were delightful women who cared for us boys with an affection and wisdom I will always appreciate.)

As I would learn from growing up in Ottawa, political leaders and their families are surrounded by people whose job it is to make life easier for them. It's one of the reasons that politicians sometimes develop a sense of entitlement. (I'm not immune myself. I once distractedly handed a buddy my coat as we arrived at a social gathering. It was draped back over my own head mere seconds later. He was a true friend.) My parents did their best to insulate my brothers and me from any assumption of special entitlement by making sure we appreciated everyone around us for the real human beings they were.

Despite her resentment over some of the strict traditions that life at 24 Sussex appeared to dictate, my mother valued many conventional, even stereotypical, aspects of being a wife and mother. She was a talented cook who often made her own bread and even, at times, delighted in performing the domestic chores of doing laundry and cleaning house. More than once, if the mood struck her, she instructed the domestic staff at 24 Sussex to take the day off and she would do their jobs herself. "I'm a nester," she would describe herself, while bemoaning the fact that few places existed where nest-building was more challenging and less appropriate than the prime minister's residence.

My father's life was strictly regimented and almost monastic, working as he did from eight thirty in the morning until six in the evening, when he would arrive home to eat dinner and spend time with his children. The balance of the evening saw him hidden away in his office, reviewing cabinet papers. Those trips to the theatre or the ballet, cultural events that my mother treasured, grew rare to the point of being almost non-existent. When they did occur, they tended to be drenched in such heavy protocol and formal obligations that much of the anticipation and joy was lost.

Once, shortly after I was born, my mother grew so desperate that she dashed out of the house with me in a stroller, leaving her security detail behind, just to spend some time free of the restrictions of being the prime minister's wife. When Pierre found out, he was both furious and fearful. Her behaviour exemplified exactly the sort of spontaneous, free-spirited, seize-the-day attitude that had first attracted him to Margaret. But the life they were leading at 24 Sussex didn't permit that sort of spontaneity. An unprotected prime minister's wife and small child were a ripe target for kidnappers, or even terrorists. My birth came just a year after the October Crisis, when the FLQ kidnapped British trade commissioner James Cross and murdered Quebec labour minister Pierre Laporte. The idea of someone seizing Margaret and me was hardly unthinkable.

MY PARENTS DEALT WITH THE COLLAPSE OF THEIR MARriage in different ways. The effect on my mother was centrifugal: the emotional impact flung her away and outward, to other countries and other people. My father turned inward, accepting in his Jesuitical way that a normal family unit was not for him. In its place he focused his monastic perfectionism on his work and his children.

As for me, I remember the bad times as a succession of painful emotional snapshots: Me walking into the library at 24 Sussex, seeing my mother in tears and hearing her talk about leaving while my father stood facing her, stern and ashen. Discovering she no longer called 24 Sussex her home. Seeing headlines in newspapers about my parents' breakup. Trying to deal with the reality and often failing.

Many children of divorced parents will say they felt guilty about the end of their parents' marriage, because they believed it was their fault that their parents couldn't live together under one roof. I don't think I ever had that guilt. I knew, even then, that the demands imposed by the life my parents were leading affected them far more than the ordinary stress of parenthood.

What I felt instead was a sense of diminished self-worth. A part of me thought I should have been reason enough for her to stay. Sometimes, hearing my parents yelling at each other, I would escape into an *Archie* comic. I would dream of growing up in mythical Riverdale, where none of the parents divorced, and where my biggest problem would be choosing between Betty and Veronica.

During this period I truly got the reading bug, and the habit has stuck with me in my adult life. Escaping into the printed page was one of the few ways I had to block out the dark drama in my parents' marriage. I quickly advanced well beyond *Archie*. Before I turned ten, I had discovered how to travel to Narnia, laugh at *Le petit Nicolas*, explore Le Guin's wizard isles, and poach pheasants with Danny, the Champion of the World. I tore through books as fast as I could get my hands on them.

At thirteen, I went to my mom and told her I wanted to read something *adult*, and she responded by giving me a copy of Margaret Mitchell's *Gone With the Wind*. I devoured it, somewhat to my father's chagrin, as we toured around the Gaspé Peninsula with my brothers the summer he left politics. From that point on, my reading tastes through my youth could best be described as eclectic, running the literary gamut. I was a sponge. I read everything from Tolkien to Tom Clancy, from la Comtesse de Ségur to Jilly Cooper, from Maurice Leblanc's *Arsène Lupin* to the cheesepulp ninja novels of Eric Van Lustbader. When my grandmother handed me Jean Auel's *The Clan of the Cave Bear* and *The Valley of Horses*, I immersed myself in a prehistoric world of discovery and adventure. From there, it was on to the classics of science fiction and sword-and-sorcery that my friends were pressing on me: Asimov, Bradbury, Heinlein, and, of course, *The Hitchhiker's Guide to the Galaxy*, whose opening paragraphs I have committed to memory. Every one of these books remains with me, lining my library shelves until my own children are old enough to learn about tesseracts, the Three Laws of Robotics, and the unique qualities of Vogon poetry.

Like many compulsive readers, I began to see the world through a narrative lens. Reading fiction alerts you to the realization that everyone around you is the hero of their own story. It's the kind of revelation that can change a young person's view of the world around them in ways they don't expect, and open their eyes to a new awareness of humanity. It affected me that way.

While travelling to the 1983 Commonwealth Heads of Government Meeting in New Delhi, my dad and I stopped in Bangladesh to inspect a dam project being built, in part, with Canadian foreign aid. On the way in from the airport with the Canadian delegation, we drove through the Bangladeshi capital city of Dhaka, where we became hopelessly snarled in traffic. I was in the back of a government car that was frozen, like the rest of the motorcade, on a main road outside one of Asia's largest and most bustling cities.

Everything and everyone around us had to wait until the traffic could move again. I looked out the side window of the car to see an older man standing with his bicycle waiting for the motorcade to move so he could cross the street. His face lined with age, he wore the weary expression of someone resigned to this kind of disruption. I remember watching him for those seconds that our paths intersected, and feeling an odd pang to realize that I would never know his story—where he had come from, where he was going, what his life was, with all the events, dreams, and anxieties that made him every bit as real and as important as I was to myself. And it struck me suddenly that he and I were just two among billions upon billions of people on this planet. Every one of us deserved to be seen as an individual, and every one of us had a story to tell.

It's not unusual, I imagine, for twelve-year-olds to have these kinds of epiphanies. Some may forget them in the next instant while others recognize that their view of life has changed within the last few moments. I had this second response. Of all the lasting memories I have of that voyage, and many other incredible trips with my father, that one— glimpsing the narrow but deep chasm between myself, the product of a privileged childhood, and the elderly man whose most valued possession may have been the rusting bicycle he had been forced to dismount—has stuck. I have never looked at my life and my circumstances in quite the same way since.

DURING THIS PERIOD I CAME TO APPRECIATE THE WIDEN-
ing difference between my personality and Sacha's. My
brother was then, as now, a faithful intellectual disciple of
my father—a man who rarely read a novel unless the author
happened to be a famous French philosopher. My father
even looked down on modern classics such as Tolkien's *Lord
of the Rings*. Such books were, in his words, "something less
than true literature." The works of Alexandre Dumas and
the detective stories of Arthur Conan Doyle were the closest things to popular fiction he ever suggested I read. Once,
when he upbraided me for reading Edgar Rice Burroughs's
Tarzan stories, I protested that the work was a classic. He
retorted that it was classic *crap*.

It may have been an act of rebellion, but I refused to
accept his view of literature. To my teenage mind, it was
beyond comprehension that anyone could be anything but
enthralled by the novels of Stephen King. Sacha and my dad
disagreed with me. To them, books like *The Stand* and the
novella *Rita Hayworth and the Shawshank Redemption* were
catalogues of things that someone had made up and nothing more. When Sach asked for books from my parents on
his birthday, his choices were encyclopedias and atlases. We
should have had an inkling, even then, that he would go on
to become a documentary filmmaker.

Sacha and I argued often about the value of different
kinds of literature, which forced me to articulate what it
was about fiction that I loved so much. I would agree that
encyclopedias could teach me facts, but only a great story

could transport me into the mind of another person. These stories taught me about empathy, about good and evil, about love and sorrow. My tastes covered many different genres, but the books I loved most proposed the idea that ordinary people (not to mention hobbits) are born with the capability to do extraordinary, even heroic things. The realization came as a sort of coda to all the lessons my parents had taught me about looking beyond wealth and appearances, and appreciating the worth of everyone I met.

It's a lesson that sticks with me to this day. No real leader can see the people around them as static creatures. If you cannot see the *potential* in the people around you, it's impossible to rouse them to great things. That may be one of the reasons why, even now, I always make time for a novel or two every month, amongst the mountains of serious works and briefing notes. Facts may fuel a leader's intellect. But literature fuels the soul.

MY MOTHER'S MENTAL HEALTH DETERIORATED AS I grew older. And there were times that I began to feel that I had to take care of her, rather than the reverse.

One day, a few years after my mother had moved out and was seeing a nice guy named Jimmy, she arrived at my school while I was in gym class saying she had to see me, she needed to talk to me, I *must* listen to her. In the school hallway she seized my shoulders and through her tears said, "Jimmy's left me! He's gone! He even took his TV!"

I did my best to console her, giving her hugs and patting her back and telling her it was all right, that things would get better. I was eleven years old.

These were painful episodes. I loved my mother as much as any child can, and seeing her in agony was as distressing as you can imagine. But it was also enlightening. It allowed me to realize that she, and parents in general, are fallible, that grown-ups aren't perfect. Within ourselves we remain children in many ways. The fears we experience early in life may be overcome with age and maturity, but they still remain, like skeletons locked in closets. During the worst stages of my mother's illness her fears and her nightmares were in many ways those of children. I don't believe she was that different from other adults her age, only that her symptoms unlocked all those closed closet doors, permitting the skeletons to stumble out and wander through her mind unescorted.

And here was another startling contrast between my parents. My mother's challenge was to deal with her emotions, and I became caught up in that process. My father's approach, which he encouraged me to practise, had little or nothing to do with emotions. It was exclusively intellectual. This was the frame he allowed for his own problems—thinking them through intellectually. At one point, he handed me a copy of Alice Miller's classic *The Drama of the Gifted Child: The Search for the True Self*.

If you're not familiar with it, the book examines children who take extraordinary steps to adapt to emotional agonies they experience. It made me realize that I had dealt with my parents' breakup by constantly seeking their approval. I had looked for ways to please them by being the good son, in hopes that this might make everything better. But it didn't, of course.

IN THE TABLOID VERSION OF EVENTS, OUR MOTHER abandoned Sacha, Michel, and me so she could turn her life into one endless party. The reality was far more complicated. My mother did not vacate our lives totally; rather, she moved in and out of our lives for an extended time. She would often stay overnight at 24 Sussex, sleeping in her old sewing room.

She and I shared a close mother-child bond, and I appreciated that she treated me in a special way—not because I was the firstborn but because she sensed that I had inherited much of her personality, including her zest for adventure, her joy in spontaneity, and her need to connect emotionally with the people around her.

Whenever I knew my mother was on her way to visit 24 Sussex, I could barely contain my excitement, and began planning my welcome. On one occasion I decided to mark her arrival with a musical theme.

I had received a small record player as a gift and enjoyed playing the hits of the day—"the day" being the early 1980s—Kim Carnes's "Bette Davis Eyes," Hall & Oates's "Private Eyes," Juice Newton's "Queen of Hearts," and especially

Journey's romantic ballad "Open Arms." (Don't laugh: that *Rock '82* album was pretty much the only non-children's music I had. But I will admit, Raffi's music has aged much better.) I had heard my mother say how much she liked the Journey song, and I decided that this would be the soundtrack to her entrance at 24 Sussex after one particularly long absence.

I waited for her to arrive in her VW Rabbit before cueing up my tiny, tinny record player in my room upstairs. As she opened the door and entered the foyer I cranked up the volume and rushed to the top of the stairs. "Listen, Mom," I yelled down to her. "It's our song!"

Her reaction was to stare up at me, happy to see me but a little confused because she couldn't hear the music at all. The volume on my record player was about half the level of a modern cell phone. I remember being crushed by that, so desperate was I to inject a sense of magic into every moment that we did have together as a family.

My mother tried to maintain the magic from time to time, with varying success. Whenever she was in New York City she would visit FAO Schwarz on Fifth Avenue and load up on great toys for all three of us. And in July 1981 she took me to London to join in the wedding celebrations for Prince Charles and Lady Diana. We stayed in the flat of her sister, my aunt Betsy, who was living in London with her husband. It was all very special, up to a point.

The evening Betsy and Robin took me to Hyde Park to watch the fireworks with a few hundred thousand other people, Mom went off to a celebrity party. The next day she described all the people she had partied with that evening: the actor Christopher Reeve, who was starring in the Superman series of movies; members of the Monty Python troupe; and, most impressive of all to me, Robin Williams.

"Oh, you really would have enjoyed it," she said to me airily, then added, "I suppose I should have brought you." It was a casual regret to her, but I, on the other hand, spent much of my youth thinking that I could have met Mork from Ork if only my mom had remembered to take me.

Eventually, my mother found her nest: a modest red brick house on Victoria Street in Ottawa, a place she could literally call her own because she had made the down payment from the proceeds of her first book, *Beyond Reason*. Sacha, Michel, and I stayed with her there on weekends and sometimes for whole weeks at a time. Free of the glamour and restrictions of 24 Sussex, she began to bloom, to reveal all the finest qualities of her personality, her intelligence, and her creativity. My father acknowledged that she had found her natural environment when she invited him over to inspect the house for himself. Stepping inside and looking around, his first response was to blurt out, "Margaret, you have . . . *a home*." It was one of those rare moments when my parents had a flash of true understanding for one another.

Every divorce has its casualties where children are involved. Our parents recognized this and, to their credit, made every effort to minimize the pain and sense of loss. They maintained a very loose form of joint custody, never arguing about the amount of time each was able to enjoy with my brothers and me. Everything involving Sacha, Michel, and me was done in our best interests. Our mother has described her relationship with our father by saying, "We didn't work as a couple, but we worked beautifully as parents."

Thanks to their efforts where we were concerned, my brothers and I never felt homesick, no matter which house we were in. Mind you, our definition of "home" began with wherever we happened to be at the time. The three of us moved as a pack, each providing the others with companionship. Along with our parents' efforts to smooth things for us, given the situation, we managed to grow up free of much of the emotional trauma that divorce can inflict on children. For that, I will always be grateful.

My mother began dating a real estate developer named Fried Kemper, and they married in 1984, the same year my father quit politics and we moved to Montreal. On their way to the courthouse on their wedding day, Mom and Fried were intercepted by my father's chauffeur bearing a large bouquet of flowers, a gesture that my mother greatly appreciated.

(The courthouse had been my father's one request: he didn't want my mom to get remarried in a church. The irony was that despite having modernized Canada's divorce laws in the 1960s, his personal faith held that "what God has joined, let no man tear asunder." He even apologized to me once, years later, for not ever being able to provide his teenage sons with a maternal presence in our lives in

Montreal: he simply felt that he could never remarry. I of course reassured him that it was of no matter to us, but the lesson he taught me about the distinction between private faith and public responsibility was one that would later guide my own thinking about leadership.)

Mom and Fried had two children, Kyle and Alicia. We three Trudeau boys played older brothers to them as they grew, having great fun in my mother's little red brick house on Victoria Street and especially at the Kemper family cottage on Newboro Lake, where life felt like one long waterfront party. Tubing along the lake, crowding around the campfire for singalongs, and hide-and-seek with flashlights in the woods were just part of the fun. As the oldest of the kids, I assumed the role of camp counsellor, organizing activities and keeping an eye on everyone, especially in the water.

As informal as it might have been, the experience was my first taste of assuming leadership and the great satisfaction of passing along knowledge and skills to others. I trace my interest in teaching and, to a degree, in politics back to those very happy, very sunny, and very memorable days.

It helped that Fried shared my father's love of the outdoors, and that he was a younger man much more in tune with my mother's fun-loving character. Dad was the guy to take us on long canoe trips, teaching us the J-stroke by urging us to practise until we got it right. Fried, on the other hand, owned a Chevrolet El Camino, a combination sports coupe and pickup truck that, as once described, you'd use for a hot date at the drive-in on the weekend and then to carry two-by-fours to the job site on Monday. My father could never relate to a vehicle like that, but it was the first car I ever drove, at fifteen, on the farm roads near the cottage. And Fried owned a speedboat, not a canoe, and a shotgun that he would use to control porcupines and other unwanted critters around the garden.

The contrast between the two men wasn't a problem for us. In fact, it was probably a blessing. With such opposite personalities and lifestyles, there was no competition between the two men. When at my mother and Fried's house, my brothers and I could unwind by watching television, playing video games, and indulging in a whole raft of other activities our father disliked. Life at the cottage and in the little house on Victoria Street was much different from living at 24 Sussex but just as wonderful. We played in the lane beside the house with neighbourhood kids and slept in bunk beds crowded into a single room. We missed our father, but we didn't miss the large bedrooms and other amenities of the prime minister's residence. On weekday mornings the school bus would arrive to pick us up and we would crowd in with the other kids. I enjoyed every minute of our visits with our mother, especially riding in the noisy school bus. Everything was totally normal—so long as you didn't look out the back window to see the RCMP security detail following us.

I remember those years fondly now. Looking back, though, I realize that I was angry. After all, most children of divorce spend more than a little time angry. But at the time, we had no clear idea what my mother was going through. The words *bipolar* and *depression* didn't mean anything to me then, and even many adults in our family were confused by the situation. My grandmother Sinclair discouraged her daughter from seeking psychological counselling because, she believed, "they always blame the patient's mother." My anger arose because it seemed that no matter how hard I tried, it wasn't enough to keep my mother near me and happy.

In recent years, as my mother has gained awareness about her own mental health challenges, we have bonded through her loving presence as grandmother to my children, and have gained closure. We talk together. We laugh together. We eat together. I take my family for weekends at her apartment in Montreal, and she comes to visit us in Ottawa. It's the relationship with her that I always wanted. I'll never be able to fix the things that went wrong with my childhood. But when it comes to spending time with the only mother I will ever have, better late than never.

The truth is, my mother was very ill. Had her illness been of the physical kind, everybody—including her family and friends—would have been more sympathetic to her and understanding of her condition. She suffered a severe mental illness at a time when such things were, at best, very poorly understood. At worst, mental illness was stigmatized and seen by many as a source of shame.

Things have changed, but not enough and not fast enough. I know, for example, what my political opponents are trying to do when they say that I am my "mother's son" more than my father's. They are appealing to those old mis-understandings and prejudices about mental illness. Like everyone, I take after each of my parents in different ways, and I am immensely proud of them both. I'm used to kind people sharing their stories with me about how my father touched or inspired them in some way, but lately more and more people approach me to say similar things about my mother. I know that her work has helped many people come to terms with their own illness or that of a loved one, friend, or co-worker. We still have a long way to go, but my mom has done a lot to ensure that people suffering from mental illness are a lot better understood than she was.

1.

3.

2.

4. My first official portrait, with my mother, Margaret.

5. I can now imagine just how my mom and dad felt as new parents.

6. Proof that those who say it was just my dad who was acrobatic with us are clearly mistaken.

7. Papa never missed an opportunity to play with us when he could fit it into his day, especially during set family time in the evenings.

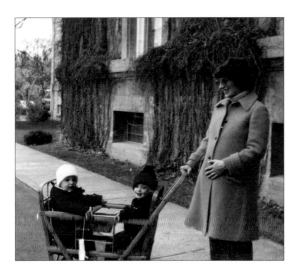

8. Mom loved getting out of the house, and the governor general's residence across the street from 24 Sussex was the perfect spot for a walk.

9. My parents couldn't resist taking a silly photo when I had the mumps.

10. Grampa Jimmy and us in our classic OshKosh B'goshes.

11. Sacha and I spent hours playing in and around the fireplace at Harrington Lake. When the fire was not lit, we loved playing with the trap door for the soot, no doubt to the dismay of our parents and the detriment of the furniture.

12. I remember thinking this bike was the greatest thing I had at the time—big shocks, a cushy seat. It was pretty much indestructible. Our bikes gave us freedom to roam the neighbourhood around 24 Sussex (with our RCMP detail in tow).

13. Sacha and I met President Ronald Reagan on one of his trips to Ottawa, in 1981.

14. Mom posing joyfully with her boys outside her new home, at 95 Victoria Street. You can see from the smile on her face how proud she was of the new house and of being with her kids.

15.

My father took me to Piazza San
Marco in Venice in 1980. I was
around the same age he was when he
had travelled there with his father
decades before.

16.

17. Another of the wonderful memories I look back
on from travelling with my
dad when he was PM. It's
hard to tell who is having
more fun in this photo as
my dad drives a tank at
CFB Lahr, West Germany,
in 1982. Also in the picture
are Lt. Jon MacIntyre of
Charlottetown (*bottom
left*) and then Toronto MP
Roy MacLaren (*top right*).

18. Seen empty the night before, the next day Red Square was packed with Soviet military and thousands of mourners for Brezhnev's funeral in 1982.

19. *Plus ça change . . .*

20. Even after my parents' divorce, we made a point of having family dinners— often at Sakura, which is still our family's go-to sushi restaurant. Here we are with Mama Ichi.

21. Mom's new family—the extended family, really—the Kempers and her three Trudeau boys.

22. The Kemper cottage up at Newboro Lake, where I spent many weekends and summers with Fried, Ally, and Kyle, is still a big part of my life.

CHAPTER TWO

Growing Up in Montreal

⸻

I SPENT MY CHILDHOOD IN OTTAWA BUT I GREW UP in Montreal. My father, my brothers, and I left the capital in 1984. It was a year of change. My father took his long walk in the snow and decided to retire from politics as soon as a new Liberal leader was chosen. I left the security of my friends and a familiar environment for a new city. My mother, who remained in Ottawa, was expecting a new baby. My brother Kyle would arrive in November.

It was also a period of intense activity by *les souverainistes* in Quebec, a back-and-forth swing between determination and despair. A few years earlier the Parti Québécois's referendum on its sovereignty-association proposal had been roundly defeated. In his concession speech, PQ leader René

Lévesque called on sovereigntists to persevere *à la prochaine fois!* (until next time), revealing that the issue remained with us. A year later the PQ won a mandate to govern Quebec with an increased share of the popular vote, confirming again that the sovereignty debate was very much alive. And in 1982, when my father succeeded in repatriating the Canadian Constitution, Mr. Lévesque called the achievement "the night of the long knives," refusing to endorse it and declaring that Quebec had been betrayed by the other provinces and, of course, by my father. In reality, Mr. Lévesque was outmanoeuvred, not betrayed, but that's not a story for this book. Meanwhile, anglophones continued to leave Quebec in droves, and language rights remained a raw issue for partisans on both sides.

In Ottawa, we had been steeped in these issues, influenced by our father's values and deep convictions. Now we were arriving to live in Montreal, our father's home, wide-eyed about the place. Throughout my life, I had spoken both languages interchangeably with my family, and almost exclusively French with my father. I was at ease with the fluidity of my French-and-English dual identity in Ottawa. With this grounding, I began my studies at Collège Jean-de-Brébeuf. It had been my father's school, known for high academic achievement, and I landed there in the midst of the political turmoil. Taken together, the abrupt new demands on my academic abilities and the strong linguistic and cultural undercurrents among students and faculty gave me a sudden new perspective on things.

My dad loved to tell the story of how he hosted his thirty-year class reunion in Ottawa shortly after he was elected prime minister of Canada. It was the height of Trudeaumania, and as the former students and teachers arrived, he was naturally proud to greet them at the door of 24 Sussex, no doubt feeling like the ultimate success story. Beaming at every old friend and teacher who walked through the door, he spotted his former science professor, who by this time was a wizened old Jesuit in the twilight of his teaching career. The professor approached my father, looked him up and down, then said matter-of-factly, "You know, Trudeau, I still think you would have had more success as a physicist."

That's the way things were at Brébeuf. Academics first, politics and everything else second. In the 1930s, the only judgment of students in the *cours classique* was where you placed in the class. Were you first? Tenth? Thirtieth? You had to be at the top if you wanted any chance at success in life. By the time I enrolled at the school as a thirteen-year-old, the culture may have become less severe, but it was still a place where parents sent their boys (no girls were permitted until the upper grades) to get a rigorous classical education. Even before you entered the main building, you knew this was a place for serious work. With its soaring Ionic columns and austere classical architecture rendered in stone, Brébeuf looked as much like a courthouse as a school. A massive stone crucifix above the main entrance signalled its Jesuit roots, although Brébeuf became non-denominational two years after my arrival.

I did well on the entrance exams. I did so well, in fact, that some school officials predicted I might match my father's legendary record as a perennial top-of-class performer. This prediction, alas, proved off the mark. The only question was whether I would enter Brébeuf in *1re secondaire* or *2e secondaire*, which were the equivalents of Grade 7 and Grade 8. Given my date of birth, and incongruities between the Ontario and Quebec school systems, it wasn't clear which would be the most appropriate choice.

Despite my father's concerns that I might be bored by the *1re secondaire* curriculum, I insisted on starting at that level for two reasons. First, enrolling in that class allowed me to enter the Latin stream, which would have been impossible if I came in at the higher level. Latin may not seem like a big draw to most people, but to me it was the language of history and adventure. Because of his own Brébeuf education, my father had been a fluent Latin speaker from his teenage years, and he used his fluency to navigate the far corners of the world on his epic backpacking expedition in the 1940s. In the Middle East and Southeast Asia, Dad's best strategy for getting information about where to eat or stay was to find the local Catholic church and speak—in Latin—with the priest.

The second, and more important, reason to start at Brébeuf in the younger grade was that I would be part of a fresh social milieu. Cliques and friendships would be established by second year, and I wasn't anxious to start my experience in such an intimidating scholastic environment as the new guy, particularly given my last name. So I began in *1ʳᵉ secondaire*, which explains why my brother Sacha and I were separated by just one grade despite being born exactly two years apart.

The students I met in my first few weeks at Brébeuf asked me a lot of questions that mystified me. Many of their questions revealed how unaware I was of Québécois slang, having grown up in Ontario French immersion and with the rather formal French spoken at home. One of the first things I was asked was, "Are you a *bollé*?" The word loosely translates as "brain." And some, upon hearing my unaccented English, accused me of being a *bloke*, to which I simply shrugged, not realizing that they were trying hard to insult me. After a few days of such taunting, I suspect they decided that I was either impervious to insults or simply mocking them right back by not reacting. The truth is that their insults and swear words were for the most part entirely unintelligible to me, and I simply hadn't the faintest idea how to respond to them.

I finally understood that although Ottawa was less than a two-hour drive from Montreal, the culture gap between the two cities was closer in distance to a light year.

The issues that inflamed many of the students were the same ones I had followed with my family from Ottawa. But this was the first time I was surrounded by people who had been living with the weight of these issues every day, and it took me a while to fully appreciate the attitudes they generated.

Sometimes things at school got personal. A few students would try to get a rise out of me by bringing up dirty laundry about my parents' separation, which had long been a staple of the tabloids. I had been somewhat insulated from this in Ottawa, both because I was well surrounded by a great group of friends who had known me since kindergarten, and because elementary-school children tend not to be as cruel and vulgar as older kids. In the Hobbesian world of high school, some kids regard anything and anyone as fair game. One day an older kid came up and thrust into my hands a notorious picture of my mother that had appeared in an adult magazine.

Hard as this may be to believe, I had never before seen that picture—never even knew of its existence. And obviously it set me reeling. But I knew this was a critical moment. If I acted shocked or hurt, it would be open season on me for the rest of high school. Everyone would know they could get a rise out of me by shoving the latest bit of gossip in my face. So I simply waved it off, leaving the bully unsatisfied, and he went off to find an easier mark.

I learned at Brébeuf not to give people the emotional response they are looking for when they attack personally. Needless to say, that skill has served me well over the years.

WHEN MOST CANADIAN PARENTS THINK ABOUT PRIVATE schools, they tend to picture small, intimate classes overseen by highly attentive instructors versed in the latest pedagogical techniques. Brébeuf wasn't like that. We students were taught in classes of thirty-six, with desks packed in a six-by-six grid, and the prevailing instructional method might be described as "sage on the stage," where the professor lectured and we wrote down what he said.

My high-school years predated the "self-esteem" movement that has swept the educational profession in recent years, in which a great deal of effort is made to help students feel good about themselves. Again, not at Brébeuf. In fact, several of the teachers seemed intent on knocking our self-esteem down a peg or two. In *4ᵉ secondaire*, or Grade 10, our French teacher, M. Daigneault, complained that students these days had no culture—and culture was like marmalade: the less you have, the more thinly you need to spread it.

M. Daigneault's course went beyond the standard curriculum to incorporate close study of thirteen works that met his high standard of classic literary excellence, including *David Copperfield*, *The Iliad*, *The Odyssey*, *Les Misérables*, and *Don Quixote*. In our first week of class, he barked at us: "Who were the Thermopylae? Come on, who can tell me? You know-nothings! Who can tell me who the Thermopylae were? I dare you!" Warily, I looked around the class. Everyone was uncomfortably staring at their desks, the floor, at anything but him. I sighed. I was going to be that guy. I slowly put up my hand.

"The Thermopylae were not a who," I said. "Thermopylae is a what. It was the mountain pass where King Leonidas and his three hundred Spartans held off the entire Persian army."

M. Daigneault nodded, pursed his lips, and resumed his rant. I managed to get some grudging praise from him that day, but really, I had an unfair advantage where that sort of knowledge was concerned because my father had engaged us in the classics from a very young age.

Years later, while a teacher in B.C., I returned to Brébeuf to visit some of my former teachers, including M. Daigneault. We had a fascinating conversation about a late-career conversion he'd had away from the rigid, intellectual, teacher-driven pedagogy he'd excelled at and imposed on us toward something much more like the more modern, student-centred approach that I had been trained in on the West Coast. Oddly, I found myself reassuring him that the rigour and excellence he had demanded and imposed had made him one of the best teachers I'd ever had, and that his demanding approach was one of the things that I strove to inject into my somewhat different teaching environment.

But however grounded I had been in the classics as a boy, I found myself tripped up on M. Daigneault's final quiz, which required each student to pick a card at random to determine which book he would be tested on. My card got me *Robinson Crusoe*. This, I remember thinking, will be a breeze. I had read Defoe's novel years earlier, like most of the books on the list, and figured I knew it well enough that I didn't need to reread it for the course. So I didn't, and, sure enough, my youthful laziness was unmasked by the professor's pointed questioning. But I did scrape through.

As we grew older as students, we chose courses that would shepherd us into streams of either arts or science. Though I had imagined myself going to law school straight from CEGEP, I wanted to keep my options open, and so I took both history and physics, which was an unusual mix. Physics in particular fascinated me, and still does, because the idea of a fundamental, primary understanding of energy and matter and how they interact appealed to me greatly.

Some assignments at Brébeuf were closely aligned with the politics of the day. One semester we held a debate on the topic of Quebec's future. The resolution set sovereignty against federalism, and the teacher thought it would be hilarious to stick young Trudeau on the separatist side. Similarly, the federalist case would be made by Christian, the class's smartest hard-core Péquiste. I cobbled together a debating position drawing from arguments I had heard over time from others, but I knew it would be difficult to argue against my own convictions. I did my best, but ultimately I felt the exercise was a success only in that it illustrated for me a truth about myself: if my heart isn't in it, I can't argue something convincingly. And my heart has always been in Canada.

In the class debate, the sovereigntists argued that independence was necessary for Quebec to reach its potential and achieve the status and dignity to which it was entitled. As the son of a proud Québécois francophone who had served as Canada's PM for more than fifteen years, and considering that another Quebecer, Brian Mulroney, was our current PM, I was at a loss to see how the province was being shortchanged. I saw no conflict between being a proud Canadian and a proud Quebecer. In fact, all I could see were the things that Quebec would give up by going it alone, everything from the Rockies to the Cabot Trail. Not to mention cutting loose the more than a million francophones living in New Brunswick, Northern Ontario, southern Manitoba, and hundreds of communities elsewhere across Canada. The compelling economic arguments against breaking up Canada at a time when the world was moving toward freer trade and more open borders clinched the deal for me. Where were the benefits? What would be the rewards? Seeing none, the whole sovereignty argument seemed extremely weak to me.

And fundamentally, even from the perspective of protecting the French language and culture, I always understood that rather than build walls to keep everything else out, it would be much better to open up, share, and radiate outwards to strengthen our identity.

But logic didn't apply. This was the 1980s, after all, and it was fashionable for young Quebecers to strike militant separatist poses, although those sympathies were not limited to students. *Indépendantistes* probably accounted for the majority of the faculty at Brébeuf. To their credit these professors avoided using their position to indoctrinate us into any particular ideology, with the exception of a history teacher named André Champagne, who did his best to convince us he was a Communist. He even kept a bust of Lenin in the corner and extolled the virtues of the USSR. But the more you probed his views, the more it became clear that much of it was an act. M. Champagne wasn't some dreamer out of the 1930s promoting a worker's paradise. He was a contrarian, challenging the settled views held by the bourgeois students passing through his classroom. His goal was to stimulate us, to get us to examine and justify what we were taking for granted in our capitalist world.

Like the other teachers at Brébeuf, M. Champagne generally hewed to the formal teaching style favoured by the school, but there were times when he liked to get into spirited verbal sparring matches with his students. He also had a habit of flinging erasers at us in a friendly way if we were nodding off, a trick I would borrow years later when I became a teacher myself. André Champagne never persuaded me to become a socialist, but he did manage to open my mind to effective strategies for challenging my own students about what they thought they believed in.

DURING MY YEARS AT BRÉBEUF I BEGAN TO THINK ABOUT language in a different manner. To sovereigntists, language was a major political issue as much as a medium of communication. You were either an anglophone or a francophone, and each label aligned you with different cultural values and perhaps different goals for Quebec. Until then I hadn't thought of myself as either a francophone or an anglophone; in my bilingual milieu in Ottawa, it simply hadn't seemed necessary to define myself one way or the other.

At Brébeuf, and in Quebec generally, the climate made me mindful of the language I chose to speak, depending on whom I was speaking to and what the subject might be. With this new awareness I began to monitor the words that popped up in my thoughts and my dreams, at times second-guessing myself as I spoke. Were the words French? Should they be in English? Decisions I had once made without thinking were becoming deliberately conscious.

I was placed in the top category in our daily English instruction classes, along with just about every other student who came from a family with at least one parent who was a native English speaker. In the eyes of some, this made us Anglos. It didn't matter if we were equally fluent in French or came from an at least partly francophone family; if you spoke unaccented English well, many kids at Brébeuf considered you an Anglo.

I felt a natural kinship with these bilingual students, so it's no coincidence that I made a lot of my best friends among this group. To these guys, the cachet of my last name rapidly wore off. Soon I was just Justin, a friend from class. Decades later, these same friends are the ones who tell me the straight, unvarnished truth. They are the people I can count on one hundred percent to tell me when I am full of it. We all need friends like that.

When I was about seventeen, we went out together for one of our first fancy meals at an upscale restaurant in downtown Montreal. Like most things seventeen-year-old boys do, this outing was organized to impress girls. I ordered *canard au vinaigre de framboise* and made a show of inhaling deeply as I emitted the words to the waiter. To this day, those friends still talk of "kicking the canard out of Justin" if I appear to be letting things go to my head. It's a great reality check. Over the years that followed, whether I was a student or teacher, camp counsellor or party leader, these good friends have always treated me the same. To them I am, and always will be, "just Justin."

I have always loved both languages, but I came to realize how very different they are, not just in the way they permit a person to express thoughts but also in the way they guide the creation of those thoughts. For example, French grammar requires you to know how your sentence is going to end before you start to speak or write, which imposes a certain rigour on your expression. If your sentence begins this way, it must end that way. This is why so many French intellectuals seem to be channelling their inner Proust even when they are speaking casually to a mass audience on television.

In English, I always felt that the grammar allows you to get to almost any conclusion, regardless of how you start your sentence. Halfway through your sentence, you can change the direction of your thought without breaking too many rules. There can be a certain sloppiness in English that is almost non-existent in proper French, where the complexity of concordance between words and within clauses requires sustained attention. Perhaps this explains why my father, who was never one to mince words on such matters, told me that he found me less persuasive in

English compared to in French. Many years later I thought about his comment when I took part in a debate that the McGill Debating Union conducted in French. Afterwards, my teammates told me I was a more formidable debater in French than in English, which, coming from anglophones, I took as a backhanded compliment.

Like many bilingual people, I sometimes flip an internal switch from one language to the other in a seemingly arbitrary way. For example, I do math only in French, because all my life, that was the language of my math classes. When I was teaching French out west, and confronted the challenges associated with getting Vancouver teenagers interested in studying a language that seemed so far from their daily lives, I used to point out the more romantic aspects of the French language. When telling someone that you miss them, you say *"Tu me manques."* So *you* is the subject of the sentence—as opposed to the English equivalent, "I miss you," in which it's all about me. It may seem a subtle difference, but hormonally charged teenagers sure got it.

MY THEORIZING ON THE LANGUAGE OF LOVE DID LITTLE to land me a girlfriend during my early years at Brébeuf. In this department, I was very much a late bloomer.

I moved to Montreal just as adolescence started to kick in, and suddenly there I was, in a city where I knew no girls, attending an all-boys school. When we finally were introduced to girls in the upper grades, it was obvious that the social habits that had made me popular with ten-year-old girls back in Ottawa were considered spectacularly uncool by girls of sixteen.

Brébeuf was an all-male school until *5ᵉ secondaire*, or Grade 11, when girls were admitted. Practically overnight, 60 girls were tossed into a class of 140 boys. At the time the policy was introduced, this must have seemed like a fine, progressive idea. But on the inside, it felt more like a sociological experiment performed by researchers intent on studying the habits and peer-group advancement strategies of teenagers.

I remember a girl named Geneviève, whom I had first met during my time at Lycée Claudel, the French school I briefly attended in Ottawa. We had been pals back then— she was not a girlfriend but certainly a girl who was a friend. Although only four years had passed since we last met, those years from twelve to sixteen marked what was probably the most significant period in our young lives of maturity and personality development. As I walked toward her, I realized that I had lost all capacity to interact with girls. The very prospect of opening my mouth suddenly seemed terrifying. I had no "game." I also made no impression on her, or at least not one that might be considered positive.

What I needed, I decided, was some unique way of establishing a social identity in this new and uncertain environment. Something that would make me stand out and show that I refused to follow the crowd. Any crowd. This led to my wearing bright green suspenders with jeans and pink flamingo ties. It was not the best decision I have ever made. My intent was to strike an ironic posture, but I never quite pulled it off. I also had a passion for nerdy showmanship, sometimes bringing in juggling balls, a magic kit, or even my unicycle to put on shows for my friends. (Yes, I owned a unicycle.) At the time, I thought this was all pretty cool. In retrospect, not so much.

It hardly helped when I developed terrible acne, something my dad had similarly endured during his own awkward teenage years. Within a few short months I went from being—or attempting to be—uninhibited to being morbidly self-conscious. The skin condition became so severe that I was prescribed Accutane, a fairly serious acne medication. My father, whose stoic nature prevented him from taking even an Aspirin in those days, was opposed to my taking the drug. This led to yet another argument between my parents. My mother eventually won this one, and I'm glad she did. It took a double course of treatments, but eventually the medication did the trick.

If I get compliments on my looks these days, I appreciate the kind words, but I always have a vague sense that people are just being polite. It's a holdover from those adolescent years at Brébeuf, and I suspect it's pretty common among folks who had a difficult time with their appearance in their teenage years.

Until the day girls were introduced into the mix, the boys at Brébeuf who earned the most respect from their peers were exceptional at either sports or academics. This made the captain of the hockey team more or less on a par with the smartest kid in the class. It was a whole new ball game with girls around. Forget your academic achievements. Now there was a high premium on athletic prowess, social graces, and comedic ability. The boys who had scored points with their brains were left on the sidelines.

Along with this change in status came a move by the boys at Brébeuf to merge into different cliques. Some boys identified with the chess club, others saw themselves as jocks, some were cool jet-set kids, and so on. My own group was composed of the bilingual kids I had bonded with in my first years at the school. My other qualities gave me limited entry into some groups. I was athletic enough to connect off the field with the jocks, and I had a big enough brain to qualify as an occasional *bollé*. I had also travelled enough to hang out with the types who skied in Europe. But my core group of friends—Marc Miller, Ian Rae, Mathieu Walker, Greg Ohayon, Allen Steverman, Navid Legendre—remained from my first years at Brébeuf.

We had no real group leader—we were just a collection of complementary personalities—but I often had a plan for some project or other. I organized us into a singing group to compete in a Brébeuf talent competition. I led us on adventures into abandoned buildings, and once even took the group on an expedition down into Ottawa's storm-drain system. From that group of strong individuals, I learned that even moments of leadership are earned through ability and ideas, and that authority is seldom conferred.

A LOT OF OUR SOCIALIZING TOOK PLACE AT MATHIEU Walker's house, on Avenue Marlowe in Montreal's Notre-Dame-de-Grâce neighbourhood. Matt's parents never seemed to mind having a bunch of teenagers in the house. Also, their kitchen tended to be well stocked with junk food, which made it even more attractive. (Ironically, Mathieu is now a cardiologist.) In later years Mathieu's home became our staging area for our forays into Montreal's nightlife.

Inviting the gang over to my house was always an option, although my father didn't usually encourage it. It wasn't all that appealing, either. Our house on Avenue des Pins was a huge, austere art deco creation that dropped down the side of the mountain from the entrance on the top floor. Directly below that was Dad's floor—off-limits to all but him—which featured his bedroom, his study, the library, and a long hallway lined with photographs and other mementoes from world leaders. Below that was our floor, and then a basement with a subterranean passage that led to the swimming pool in an annex. Adding to the somewhat less than kid-friendly atmosphere, my father imposed rules about the language to be spoken on each floor. The very top floor, for instance, was exclusively French. So if he heard my friends and me there speaking English in the kitchen or living room, we could expect a reprimand. Having lived with that somewhat arbitrary discipline all my life, I found nothing especially strange about it. But to my friends, it was truly odd.

On our floor, Sacha, Michel, and I had our separate bedrooms and a family room. This floor of the house was always loud with banter, horseplay, sibling disagreements, and a lot of laughs—basically boys being boys. As much as we were developing into different people, the bond between us was solid, and we supported each other through those teenage years. But it didn't stop my friends from being surprised by what went down in the family room. Our family room featured overstuffed low couches and tumbling mats that Dad had bought to encourage our regular pastime: play-fighting. We were raised in judo, and so did a lot of grappling. But then we'd grab sticks and wooden swords and just go at each other in a more or less controlled fashion. There were few rules, other than no punches in the face and no biting, and if someone got hurt, we stopped. The first time my friends saw us wrestle, they were shocked by the intensity of our bouts. As I said, by this time in life, it was clear that Sacha, Michel, and I had very different, and sometimes conflicting, personalities. Throwing down with each other had always been our way of working through the rough justice of the nursery.

There were times when our arguments really did get out of control. I remember one time we drove our father's Volvo down to our mother's cottage at Newboro Lake, with me at the wheel. I was about eighteen, Sacha sixteen, and Michel fourteen. For some bizarre reason, we got into a raging argument about who would control the car windows. It was the kind of thing that only teenagers could get riled up about, but it got so heated that I pulled over to the side of the road and we all piled out of the car to have a real, not play, fight. Michel and Sacha teamed up to pin me to the ground, there was much grumbling and many insults, and eventually we cooled off enough to get back in the car and tolerate each other for the rest of the trip.

When we returned to Montreal and our father heard about the incident, he read us the riot act. "No matter what happens, the three of you need to stick together," he told us, adding that he would not tolerate hearing anything like it again in the future. For his sake, we did our best to avoid another roadside fight.

Only now that I have children of my own do I realize how painful it is for a parent to see his children fight with one another, and why our skirmish upset him so much.

I BROUGHT MY LIFELONG LOVE OF ATHLETICS TO SPORT at Brébeuf, playing lacrosse and regular pickup games of full-contact football. I even had a brief stint on the gymnastics squad. Still, at Brébeuf, as in Canada generally, hockey was king. You might think that my father, the archetypal Canadian outdoorsman, would have encouraged us to throw on some pads and grab a stick while we were still in kindergarten. But that wasn't the case. From the beginning, he had emphasized the importance of testing yourself individually, seeing how much you could do and how far you could go on your own, not counting on others to bail you out. He had also decided that he would not spend his early-morning hours shivering at a rink watching a bunch of peewee players

ankle-skate up and down the ice. Many of my friends' families chose hockey, but we chose cross-country and downhill
skiing according to my father's wishes, and because it was
something we could do as a family in the outdoors.

I also suspect that the somewhat arbitrary rules of hockey
and other sports grated against his universalist sensibilities.
It was more important to him, as an outdoorsman, to follow
the immutable rules of nature rather than those imposed
by a man with a whistle and wearing a striped shirt. When
we learned to skate it was on the Rideau Canal, with nary a
puck or hockey stick in sight.

Things grew worse when I showed up at school with
my hockey equipment. Having the correct—or fashionable—brand of stick is a very important mark of being
cool among Canadian high-school students. I had hoped
my father would take me to a sporting goods store where
I could buy the brand that would assure my admission into
my peer group, even if my hockey skills wouldn't. Instead,
he took me to the storage room and pulled out a strange
blue piece of wood that he had received on a state visit to
Czechoslovakia several years earlier. He assured me it was a
hockey stick, and a good one too. I had my doubts. I don't
recall the brand, only that the name was unpronounceable
and included a variety of obscure accents above the letters.
This would definitely not make me cool.

In the school rink I felt like the kid from Roch Carrier's
story "The Hockey Sweater." Instead of wearing the wrong
jersey, I was wielding the wrong stick. My friends took one

look at that stick and instantly knew I wasn't going to make the school team.

I BENEFITED GREATLY FROM THE RIGOUR AND DEMANDS of the curriculum at Brébeuf and from the environment I was raised in. But despite the fine education, like many teenage boys, I applied myself unevenly. I worked hard at the classes I liked, and went through the motions with the subjects that didn't appeal to me. When I was bored, I would open a novel in my lap and escape the tedium of the classroom. I was always confident that I would be able to score a decent mark at exam time, and usually I did. But I was coasting through school. My teachers and I knew it.

One day, my math teacher decided enough was enough. After watching me drift through a half-hearted performance on a series of assignments, she called me into her office and sat me down for a serious talk. "Justin," she said, "I have watched you slide your way through every class at this school. You're smart enough to pull it off. But you're not putting enough work into your studies."

I began to tune out. I had heard variations on this speech several times, so the words didn't make much of an impression on me. Then she dropped the bomb.

"Do you know what I think?" She paused, knowing her next words would bite. "I think maybe you believe that you don't need to work hard because of who your father is."

I was fifteen years old and in Grade 9, and this was the

first time anyone had said such a thing to me. I'm sure some teachers had *thought* it, dancing around the issue when they were calling me out for being lazy. But no one had flat out accused me of trying to coast through school on my family name. Angrily, I blurted out, "Don't be ridiculous."

Our father had always been careful to ingrain in us the principle that the Trudeau name was not a currency to be spent but a badge of responsibility to be worn. If the teacher had accused me of doing *discredit* to the Trudeau name, I might have accepted it as a reasonable claim. My father himself had told me several times that he was disappointed with my average academic performance at Brébeuf. But to suggest that I expected special treatment because of who my father was, well, that was simply wrong.

Yet the more I thought about her comment, the more I realized that her words were significant, even if I deeply disagreed with their premise. They made me understand that even if I wasn't trying to trade on my family name, it wasn't unnatural for people to *suspect* me of doing so. She had voiced a presumption that I knew I couldn't ignore if I didn't at least try to rise to my potential.

I went on to graduate from Brébeuf and earn two university degrees. During much of this period, my academic inconsistency was a source of concern not just to my teachers but to me as well. Eventually I came to understand that the problem was rooted in something more serious than a temporary bout of adolescent laziness. When I fairly deliberately flunked Experimental Psychology (the course's emphasis seemed almost entirely to be on how to produce properly standard-ized lab reports, which annoyed me), it was a wake-up call that I had issues I needed to deal with.

It wasn't an attention deficit, because I was more than capable of sustained focus on a subject when I felt like it. In fact, I could become so enthused about courses I enjoyed that I often became a sort of informal teacher's assistant, helping other students with their problems.

Failing the Experimental Psych class led to a serious heart-to-heart with my father in his study at Avenue des Pins, where something extraordinary happened: I realized, and announced to him, that I was not like him. All my child-hood, my father had been my hero, my model, my guide, my instruction booklet to life. But when, trying to be help-ful, he showed me his report cards dating from his time at Brébeuf in the 1930s, featuring a straight line of As stretch-ing from top to bottom, I knew we were fundamentally dif-ferent people, with different approaches to life.

He was proud of the approval, the recognition; he was driven to show how smart he was, and he worked diligently at it. In contrast, I had always told myself that if I worked hard, it would be for a greater purpose than just a grade on a paper or plaudits from authority. I rebelled against jumping through hoops for their own sake and resented artificial competition.

But at the same time, I knew that I did face a real chal-lenge: I was stuck with a mildly crippling form of perfection-ism, as in the phrase "the perfect is the enemy of the good." A light panic gripped me whenever I sat staring at a blank page, preparing to start an assignment. Embarking on an essay that I knew would never match others' (let alone my own) expectations filled me with anxiety. Which, in turn, created a sort of subconscious defence mechanism that worked like this: if I choose not to give a project all of my effort, I cannot be judged negatively on the basis of the result. My father strove for and succeeded in reaching tremendous heights of achievement. I chose not to try nearly as hard, so why should anyone be surprised when my marks failed to match his?

Flunking Experimental Psych effectively killed any hope I had of going to McGill Law straight from CEGEP, which was what the best and the brightest did. I had sabotaged that path, perhaps as a way of forcing myself, and my father, to come to grips with the fact that I would never be the academic high achiever he was. That path was not mine.

I knew I was more than capable. Whenever I took a high-stakes one-off standardized test, such as the ones administered by Brébeuf to new applicants, the results were top-notch. On the SATs I took in my last year of high school, I scored 1400, putting me in the top 5 percent. This was good enough to get me into arts at McGill, despite my erratic grades. A few years later, when, mostly on a lark, I sat for the Law School Admission Test, I cleared the ninety-eighth percentile. So I knew I was smart: I just needed to find my own path. Which is why I chose to study literature. I would apply my intellect to something I was truly passionate about—reading—and give myself some time and some tools to understand myself better.

WHEN I STARTED AT MCGILL UNIVERSITY IN THE FALL of 1991, my friends from Brébeuf remained at the centre of my social life, but I managed to make some new friends on campus, one of whom remains particularly important to this day.

I had been at McGill for just a week when I ran into Jonathan Ablett. Jon and I had gone to the same elementary school back in Ottawa and we encountered each other on the steps of the Shatner Building, the heart of the campus where most of the big student groups gathered. After we caught up on each other's lives, Jon asked me if I had made many new friends at McGill. I shrugged, and offered that since I was a Montrealer, I already had lots of friends around and hadn't really looked for more. The truth was, I didn't know how I was going to make new friends, and wasn't sure I wanted to try. Jon glanced around, beckoned to a long-haired guy standing nearby, and introduced him as Gerry Butts, vice-president of the McGill Debating Union. Today, almost twenty-five years later, Gerald is not just still a best friend; he is my closest advisor as leader of the Liberal Party of Canada.

On Gerry's invitation, I joined the Debating Union, where we became fast friends and I spent the next year honing my skills and travelling to tournaments. It was an education on its own, focusing my ability to think on my feet, to spot a weakness in an opponent's argument and exploit it with the right combination of logic and turn of phrase.

I also learned that debating at the college level is as

much about the quick-witted ability of a stand-up comic as it is about logic and fine rhetoric. This was especially true whenever the resolution being debated was some frivolous subject such as whether baths are better than showers or whether winter is better than summer. Given those kinds of topics, the most successful debaters are gifted comedians. It took me a while to fully appreciate this, because my sense of humour is more on the wry side than the kind that generates belly laughs. Eventually, though, I learned the ropes and adjusted my delivery.

Debating also provided me with an interesting window into some of the important issues reverberating around university campuses in the early 1990s. Many of the outstanding McGill debaters were women who were active feminists. I remember going toe-to-toe over beers with a few of them on the issue of whether a man could be a feminist. Some argued that by definition alone, feminism demanded a female perspective, while I suggested that the exclusion of men was antithetical to the egalitarian principle at the core of feminist thought.

As you might imagine, there was significant overlap between the women in debating and both the McGill Women's Union and the Sexual Assault Centre of McGill's Students' Society. When, in my second year, the centre began recruiting male facilitators, one such friend, Mary-Margaret Jones, encouraged me to get involved. Women's issues had come to the fore for me with the horrific massacre at the University of Montreal's École Polytechnique a few years before, which happened a stone's throw from my high school. I had also grown weary of debating on either side of any issue: I wanted to use my communications skills in the service of something meaningful.

In addition to its crisis hotline, the Sexual Assault Centre had created an outreach group to meet with students at fraternities and residences, and I was part of the first cadre of men trained to join the women activists in leading discussion groups on sexual assault and date rape. We used role-playing exercises and other interactive methods to start students thinking about sexual assault in a different way. This new perspective was important, because many people believed that rape was something that happened when a stranger jumped out of the bushes. We wanted everyone to understand that the vast majority of sexual assaults are committed by people known to the victim, and are as much about power as they are about sex. We suggested communications methods that women could use to handle situations before they became violent and coercive, and we taught men how to recognize the messages women were sending them. It isn't just "No" that means no. "I'm not feeling comfortable with this" also means no, as does "Maybe we should go back to the party."

I like to think that our work at the Sexual Assault Centre began to have results, at least for the students if not for the institution. When McGill's administration made a somewhat controversial staffing choice for the newly created post of sexual-assault ombudsperson, another student and I spoke with the university's president about our concerns. It was a lesson for me on how resistant to dealing with delicate issues institutions can be: we were thanked for voicing our perspective and politely ignored.

NOT SURPRISINGLY, MY YEARS AT McGILL WERE A TIME of major social transformation for me. I became less gangly and I felt more confident about my appearance, banishing some of my lingering teenage insecurities. I was still living at home with my father, who gave me no small measure of freedom, and I immersed myself in the pleasures and perils of an adult social life.

Until I was eighteen, about the only alcohol I had sampled was the odd glass of wine during dinner. My choice not to drink during high school qualified me as the designated driver at Brébeuf parties, the guy who said, "I don't need alcohol to have a good time," prompting other kids to roll their eyes. But I meant it, and I still do. With a few rare exceptions, notably the months surrounding my father's death, over-consumption of alcohol has never been an issue in my life. A nice cold beer from time to time, a glass of wine with a good meal: I simply don't drink much.

That said, I did go through a brief partying phase while at university. Friends had rented a fabulous apartment on Rue Émery, just around the corner from Théâtre St-Denis, and we had some great parties there. One evening, I became tipsy enough to put on the costume of the mascot for the

McGill Martlets, the name given to some of the university's athletic teams, and run out into the street. (Don't ask how it had ended up in that apartment.) The mascot looks like an angry red swallow, and I decided that the peak of hilarity would be for that giant angry swallow to smack itself into the windows lining the street, startling the Saint-Denis café-goers. Suffice to say, this did not go over well, so my friends caught up with me, promised we were all going to an even *better* party, gently removed the slightly battered layers of plush, hailed a cab, stuffed me alone into the back seat, and gave the driver my home address.

When I arrived home, my father was returning from a dinner, and he was most unimpressed with the state I was in. The next morning he gave me a stern talking-to about the perils of alcohol, which I sat through grimly, not bothering to explain that of his three sons I was the most strait-laced and he had nothing to worry about. But I was in no state to argue.

I WAS DATING MY FIRST SERIOUS GIRLFRIEND AROUND this time. We had started going out back at Brébeuf, and the two of us were still hanging out with our old crowd. With all the new people I was meeting, it was a relief to sometimes not have to think about my last name and the effect it had on people when I first met them. I'll admit that sometimes in those initial meetings I deliberately left out my last name. Ideally, I didn't want them to hear "Trudeau" until I'd first had time to make a strong impression with my personality.

Sometimes this demanded a little improvisation on my part. My girlfriend was at Concordia, and I accompanied her to their debating club's recruiting night even though I attended McGill. We were arguing a resolution that the city's Olympic Stadium should be torn down and the materials used to build a bridge from Quebec to Newfoundland. When it was my turn to get up to contribute, I gave my name as Jason Tremblay. I felt a bit of a thrill: no one knew me, there were no consequences because I wasn't joining the Concordia team in any case, and I got to speak without any expectations from anyone of how good I should be. For everyone in the room, I was a completely blank slate. Perhaps because of that absence of pressure, I argued quite well, suggesting light-heartedly that the bridge was a great idea, since blocking the current through the Strait of Belle Isle would surely divert the North Atlantic Gulf Stream toward Canada, resulting in a more Mediterranean climate on our east coast. "Just think about it," I declaimed. "Olive oil from Nova Scotia!"

At the end of the meeting, the organizers asked me to join the squad. I shook my head ruefully: "I don't go to Concordia," I admitted, "and by the way, my name's not Jason Tremblay." I never wanted to hide my identity, but for a moment it was nice to step away from it.

The only other time I have ever given a false name was a few years later, when I started boxing at the east-end Club de Boxe Champion. It's a bit intimidating at the best of times to walk into a real boxing gym and sign up, and this gym was a tough place, where being a former prime minister's son wouldn't have made me very popular, except perhaps as a punching bag. So I made it slightly easier for myself by signing in with a twist on my mother's maiden name, Justin St-Clair. I wanted to be known first by my work ethic and skill, not by my parentage.

After about a year, my coach, Sylvain Gagnon, told me he'd figured out my real name, but by then I was established as a serious member of the gym community and it no longer mattered. And that was how I liked it: have people get to know me first, then my last name didn't matter so much later.

Sometimes people were simply fascinated by the cachet of my last name and they would try to attach themselves to my social circle for the wrong reasons. I became attuned to that sort of thing over the years and developed a social sixth sense that continues to serve me today.

Whether the reaction of others to my name was good or bad, I didn't like the idea that people would have pre-conceived expectations of me before they heard what I had to say in a debate or saw what I could deliver in a box-ing ring. In either location, some of my opponents would either pull their punches or go out of their way to pummel me. I'd also learned that my natural caution was useful in all situations. With people I met in social situations, my instinct was to broadcast a strength of personality that would either define me before they knew my last name, or override (or at least mitigate) any preconceptions they had if they already knew it.

Of course I wasn't the only one dealing with the challenge of what it meant to be "a Trudeau." It affected my brothers as much as it affected me, and because each of our attitudes toward our last name tended to reflect our relationship with our father, it brought out the personality differences between us. Sacha, who most faithfully strove to emulate our father's example as an intellectual and ascetic, strengthened his defences and protected his privacy. Michel, in contrast, rebelled against my father's influence and did his best to live life in almost complete anonymity, first going by Mike at summer camp, heading east to Dalhousie for his undergrad, and ultimately choosing the West as his home. I occupied the middle ground. My Trudeau identity was a source of great pride to me, but I also wanted to be judged on my own merits, as someone whose emotional temperament and intellectual attitudes stood apart from my father's.

On occasion, my background and family name have led to incidents that were comic and surreal. Like the day, during a trip to Paris, I struck up a conversation on Boulevard Saint-Michel with a retired American professor who had made a name for himself translating Robert Frost's poetry into French. He was an interesting and eminent character who, when I mentioned I was from Canada, began rhapsodizing about "that wonderful prime minister you had in the seventies, the one with the beautiful wife who ran away."

I couldn't resist. I said, "You mean *Mom*?"

An even more hilarious incident took place in 1992, when my father and I went on an eight-day rafting trip down the Tatshenshini River in Yukon and northern B.C. Our purpose for the journey was to help raise awareness of the potential environmental dangers to the region posed by a copper mine.

Dad and I were to meet in Whitehorse. I arrived a few days ahead, planning to do a little sightseeing on my own. By sheer social happenstance I met and made fast friends with a group of bikers in town for a summer motorcycle rally. They were all good guys, although they had the appearance you would expect of rough-and-ready bikers willing to ride thousands of kilometres on two wheels in the open air just for the joy of it. I grew closest to a biker named Big John, who owned a Harley-Davidson dealership near Pittsburgh. I didn't tell Big John and his buddies my last name. As Americans, I thought, they likely wouldn't have known it or cared anyway.

My father arrived in town a few days later, and when he and I entered a crowded Whitehorse bar, I heard a familiar booming voice shout from a far corner. "Hey, YOU! We don't want *your* kind in this place." It was Big John bellowing at us from his table in mock anger, and I could see him grinning at his own joke. My father, however, had become stiff as a board, imagining we were about to be verbally assaulted (or worse) by some long-disgruntled voter nursing a decades-old grudge.

I took him over to meet Big John and his buddies, and

when he realized they had been shouting at me, my father took the situation in good humour. He could very graciously play the straight man.

When we returned home, Dad told my mother some details about the trip and added, "You know, I never realized it, but Justin is really very good with people."

DURING MY UNDERGRADUATE YEARS IN ENGLISH LITER-ature, I read hundreds of books, wrote many essays about writers as varied as William Blake, Aldous Huxley, and Wallace Stevens, and generally experienced the broad intellectual horizons that are the goal of a liberal arts education.

It was also a time for me to try on and often discard all manner of political postures and ideologies. This kind of thing happens to men and women in their late teens and early twenties who arrive on campus in an idealistic state of mind. Soon they are seeking answers to big, dramatic questions like, What is the meaning of life? How do we build a better society? What is standing in the way of social justice? Their search steers a lot of campus intellectuals in the direction of totalizing ideologies, such as dogmatic Marxism or Ayn Rand's theory of objectivism.

I was as curious about these questions as any other college student, but I always was suspicious of cultish, reductionist movements. My father was fond of quoting Thomas Aquinas's admonition *hominem unius libri timeo* (I fear the man of a single book). I internalized that: whenever a classmate or friend tried to convince me that the answers to life's big questions or major political issues could be derived from *The Communist Manifesto* or *Atlas Shrugged* or some other single-minded philosophy, I grew wary. One of the lessons of life I learned from my father was that the world is too complicated to be stuffed into a single overarching ideology. I was exposed to all sorts of political influences on campus, but when I graduated I was the same open-minded centrist I had been when I first arrived.

WHAT GREW MOST DURING MY UNIVERSITY YEARS WAS my own understanding of Quebec, federalism, and the nature of Canada generally.

I had heard my father describe the political atmosphere that existed in Quebec during his youth and was struck by the many differences between his time and my own. During the 1940s and '50s, Quebec nationalism had been a power-ful force, tied not to separatism in the way we think of that term today but to something much different. In my father's youth, Quebec's political and religious elite were concerned about protecting the province's French Catholic character within largely Protestant North America. Simply put, the emphasis was therefore on maintaining a society of farmers and lumberjacks, with a small cadre of lawyers, priests, doc-tors, and politicians to oversee it. Money and business were left to *les Anglais*.

This situation grew untenable, of course, by the mid-twentieth century, and a number of thinkers, artists, and writers (of whom my father was one) fomented the Quiet Revolution, making education, urbanization, and secularism key pillars of modern Quebec.

Quebec began to assert itself more fully, and the nationalism that gained strength in the 1970s, and which I experienced during my years at Brébeuf in the 1980s, usually expressed itself in demands for more governmental powers and greater recognition of the province's linguistic and cultural character. The federal government's patriation of the Canadian Constitution in 1982 without Quebec's explicit sign-off led to a decade-long mobilization to find some new formula that would reapportion powers and satisfy Quebec's concerns. Politicians and constitutional lawyers began searching for some kind of grand bargain, and the result was the failed Meech Lake Accord of 1987.

The referendum campaign surrounding the Charlottetown Accord of 1992, which coincided with my second year at McGill, scaled my engagement with Canadian politics.

A handful of Canadian federalists, including my father, opposed the accord largely because it appeared to signal a capitulation to Quebec's escalating demands on Ottawa. Section 1 of the accord would have amended the Canadian Constitution to stipulate that "Quebec constitutes within Canada a distinct society." It also would have declared that "the role of the legislature and Government of Quebec to preserve and promote the distinct society of

Quebec is affirmed." Under section 21, Quebec would be guaranteed no fewer than 25 percent of the seats in the House of Commons, no matter how future populations might shift.

I had always defined myself as a Canadian federalist. How could I not? But in the early 1990s that label wasn't enough, because the various reform proposals being tossed around forced us all to think about what an ideal sort of federalist structure should look like. As the debate about the Charlottetown Accord went on, I began to study the document closely. By the time I finished marking up almost every page with a highlighter, I realized that the problems with the document extended beyond Quebec. The accord contained a long list of concessions to the provinces generally, with very little coming back to the federal government in return. For me, this was the larger issue. I am not, and never have been, one of those federalists who believes Ottawa must involve itself in every area of policy. But the Charlottetown Accord would have tipped the balance too much toward decentralization, locking in federal funding for all sorts of programs while limiting the federal government's ability to impose national standards. Something was wrong here.

Take a couple of examples that bugged me: Section 38 dictated that the Canadian government would agree to all but abandon its power of disallowance, which permits the federal government to overrule a provincial law when it conflicts with national goals. And under section 39, Ottawa would have lost its "declaratory" power to classify certain areas, such as the control of a vital resource, within the federal government's domain. There was nothing wrong with proposing these concessions, and they were certainly defensible from the point of view of the provinces, certainly since neither had been much used in recent times. But I kept returning to the same question wherever these and similar concessions by the federal government were addressed: What was Ottawa getting back in return? That side of the ledger seemed mostly empty to me.

Just to add frustration to my concerns, I discovered that many of the people who said they supported the accord admitted that they hadn't actually read the thing, certainly not in the detail that I had and that I believe it demanded. I specifically remember accepting a pro-accord pamphlet from one of McGill's Young Progressive Conservatives, who were actively seeking its support on campus. After reading the pamphlet over, I asked the YPC activist why he and his friends weren't also providing students with copies of the actual accord. He brushed me off by claiming that the bullet points in the pamphlet were all that people really needed to know. Why should they read the whole complex document, he suggested, when we've gone to the trouble of underlining its key points for them? Because, I replied, people were taking a strong stand on the future of our country without educating themselves about what that future would look like. And it couldn't be summed up in a dozen or so bullet points.

I made a real pest of myself that year at McGill, going around campus with my dog-eared copy of the Charlottetown Accord, lecturing friends about this or that provision. I wish I could say that I changed a lot of minds, but the truth is that most people took positions according to their existing political affiliation. Supporters of Brian Mulroney's Progressive Conservatives tended to support the accord, while his critics tended toward skepticism. The pro and con sides splintered in various ways. Eastern federalists supported the accord, but followers of Preston Manning and his Reform Party rejected it because of its deferential treatment of Quebec. (They weren't happy with the scope of the accord's Senate-reform provisions either, proving there's not much new in our politics.)

The separatists led their own charge against Charlottetown, which caused some people to misinterpret the nature of my position. What did it mean, they wondered, if both the Péquistes and I were opposed to the same concept? Did I have anything else in common with separatists? I remember a call-in radio-show producer hanging up on me because he refused to believe that a self-described federalist who was opposed to the accord could add anything to the debate. In desperation I took to wearing a T-shirt that read *My No is a federalist No.*

In the end, the Charlottetown Accord was defeated in the October 1992 referendum by 54 percent to 46 percent. In Quebec, it lost 57 to 43. All four Western provinces voted against it. I was pleased with the result. I was also a little intoxicated by the experience of diving into an important political issue, marshalling the best arguments on behalf of my position, and becoming a passionate advocate for it. The episode sharpened my feelings about Canada and about protecting the things that make it strong, distinctive, and politically coherent. The months I spent carrying around that well-worn copy of the Charlottetown Accord, prepared to discuss its defects with anyone who wanted to engage me in a debate, marked an important step in my journey toward political life.

Three years later, I was consumed, as we all were, by another political campaign. This time, the stakes were higher than mere amendments to the Constitution: this time it dealt directly with the possible dissolution of the country.

It was October 1995, and Quebecers were set to vote in the province's second referendum. If the Yes side succeeded, the province would be primed, backed by a majority of its citizens, to begin negotiations disassociating itself from the rest of the land. When polls taken barely a week before the referendum suggested the separatists could pull it off, many of my Montreal friends and I feared we were living through Canada's final days in its existing form.

I remember feeling outraged throughout that campaign at how the Yes side used propaganda and demagoguery to try to sell their idea. It seemed to me that they actually didn't understand the seriousness of what they were proposing. If you're going to create a new country, you should have clear support and desire from the population to do it. You shouldn't have to trick them into it, or sugar-coat it, because the inevitable challenges that would arise during the transition stage would require continued public support. Given the certain growing pains of any new country, people would have to feel that it was all worth it. And what a possible Yes victory seemed to look like—a bare majority mandate based on misinformation—struck me as a recipe for upheaval and unrest.

With three days to go, my friend Ian Rae and I joined an estimated hundred thousand people in downtown Montreal for the Unity Rally, an event that remains, to this day, the biggest single political gathering in Canadian history. Giant Maple Leaf flags flew everywhere, and Place du Canada was mobbed with supporters of the No side. Wanting the best possible view, Ian and I headed to the nearby CIBC skyscraper and climbed some scaffolding onto the building's second-floor lobby terrace. If you look at the famous giant poster print made from an overhead shot of that rally, you can see us near the twin white media tents atop that CIBC terrace. To be surrounded by so many Canadians was a stirring experience, and it helped soothe my jangled federalist nerves.

On referendum night, my brothers and I watched the returns at home with my father. (He had finally relented in his opposition to a television, thank goodness.) The No side won by the narrowest of margins, a mere 50.58 percent to 49.42 percent in favour of the federalist position, a difference of just 54,288 votes. Through it all, my father remained oddly unfazed, and when the official result was announced he nodded, said "Good," and calmly retired to bed.

But hey, this was something to celebrate. I met my friends at a bar on Rue Metcalfe, where we heard rumours that separatist mobs were planning to invade the downtown from the direction of Parc Maisonneuve to the east. The rumours proved baseless. If any protests had been contemplated by the Péquistes and their supporters, they were probably discouraged by the sight of riot police crowding the downtown area. Their ominous presence reinforced our sense that a disastrous result had been narrowly averted that night.

ALL THESE YEARS LATER, I STILL THINK BACK TO THAT day from time to time and imagine how much our country would have changed if a mere 27,145 No voters had decided to cast their lot with the separatists. Canada would probably no longer exist. And what message would we have offered the world? If even a country as respectful of its diversities as ours had failed to reconcile its differences, what hope would the rest of the world have of getting along?

To this day, that question is one that drives me.

CHAPTER THREE

Travelling East, Going West

—

M Y GRADUATION FROM McGILL IN 1994 deposited me at a crossroads. I was twenty-two years old, with a freshly minted BA in English literature. My university years had included a few of the same academic problems I had encountered in my years at Brébeuf, but my undergrad marks were good enough for me to have options for my next steps.

I had chosen to do my undergraduate degree in literature not just because of my love of reading but also because it ensured I would continue my studies. While it made for a great first degree, for me it simply couldn't be a last degree. The challenge was, I didn't yet know in which direction I wanted to go.

Perhaps foreseeing this challenge, a number of my old Brébeuf friends and I had planned a big trip for the year after graduation. Up until that point, I had travelled to more than fifty countries around the world, mostly with my father, but here was a chance to build on that. I packed a few things into my backpack—it's remarkable how little you actually need to pack when you realize that it's impossible to bring all you'd need for four seasons across three continents—and headed across the Atlantic.

I spent the summer in France, mostly on my own, travelling from Provence to Normandy and finally settling in Paris, where I spent most of my days in museums and libraries. Having removed myself from everyone and everything that made up my normal surroundings, and struggling with a shyness that prevented me from making friends easily, I found myself with a lot of time to think about my life and my future.

I thought about my father's path at my age: fierce academic achievement at Brébeuf, followed by top place at the University of Montreal's law school, then a master's at Harvard, followed by more studies, but no degrees, at the London School of Economics and the Sorbonne. He then spent many years in a wide range of pursuits— travelling around the world, working for a short stint as a lawyer, publishing a subversive intellectual magazine that contributed to Quebec's Quiet Revolution, writing a book or two, teaching constitutional law for a few years— before running for political office in his mid-forties. I had already unhooked from that track, and my self-examination confirmed that a meandering path as a public intellectual was not for me.

My mother had got a sociology degree from the brand-new Simon Fraser University in Vancouver, then moved east to get married and start a family with my father. And although I already knew that I wanted a family, I wanted to be neither as old as my father nor as young as my mother when I began one.

That summer, in a quiet moment of reflection on a hillside, I realized my next step: I would become a schoolteacher. This would be my way of having a positive influence in the world. The job drew together my strengths for learning, for sharing, for understanding people. And importantly for me at the time, it was all mine: it would be my way of freeing myself from my family and our past.

I excitedly called home to share my epiphany. "Justin, that's wonderful," my mother said. "You know you come from a long line of schoolteachers back in Scotland."

Oh well, I thought, at least it would be a break from my family's recent past. With the plan in place to begin the following fall at McGill's Faculty of Education, I was ready to focus on the year of travelling ahead of me.

I joined three of my best friends in London in September, and together Mathieu Walker, Allen Steverman, Marc Miller, and I embarked upon a tremendous adventure. We joined a motley group of Brits, with a few Aussies and a lone Finn, bound for Africa on an overland truck expedition. We sped across France and Spain in a few days, camping behind highway rest stops, impatient to get off European soil. We made our last phone calls home from Gibraltar and loaded onto a ferry to Morocco.

Morocco was medinas in Fez and Marrakech, hikes through the Atlas Mountains, and picking mussels for breakfast off the rocks in Western Sahara where the desert meets the Atlantic. We then crossed an empty stretch of the Sahara into Mauritania, where my memories are of having to push the truck over sand dunes, becoming terribly sick after eating leftover tuna salad, delighting in randomly having delicious Korean food at a fisherman's house in a small village, and unsuccessfully hiding our last cases of beer from customs agents.

The transition from North Africa into West was a welcome one. Mali was mostly friendly and diverse, but with an edge. Matt got mugged and pepper-sprayed in Bamako, losing a few dollars but not much else; Marc beat the village strongman in an arm wrestle after an archaeological trek through the ruins of an ancient civilization; and we visited a near-abandoned community where we were shown a tree under which, we were told, within living memory children had been sacrificed as part of religious ceremonies.

Onward through Burkina Faso and into Ivory Coast, then Ghana, Togo, and Benin. Again, more contrasts: beautiful places and friendly people, interspersed with wrongs both historical and current, from slavers' forts from which millions of Africans were sent to lives of bondage on the other side of the Atlantic, to current excesses like an empty cathedral bigger than Saint Peter's and a presidential palace complete with crocodiles in a moat.

It was late December when we reached the Nigerian border, and time to move on to the next phase of our trip. We caught an Aeroflot flight out of Cotonou to Helsinki via Malta and Moscow. Marc then headed home to Montreal while the remaining three of us applied for our Chinese tourist visas at the local embassy while staying in a one-room apartment belonging to the aunt of our Finnish travel friend. We spent Christmas that year in Helsinki, but by New Year's we were hurtling across the steppes on the Trans-Siberian Express.

It was another unforgettable experience, notwithstanding the appalling food and service aboard the train. The USSR, never known for either the appeal of its cuisine or the quality of its customer service, had been dead for a few years by then, yet the idea of satisfying consumer expectations obviously was still an alien concept.

The timing of our trip meant that the train was filled with Chinese students returning from Russian universities to celebrate their country's New Year's holiday. I spent much of the week enjoying the landscape, drawing sketches, and, appropriately, reading *War and Peace*. On New Year's Eve, the train's conductor, sensing an opportunity to practise his quite serviceable English, invited us to join him in consuming large amounts of vodka and discussing the state of the world. If the stories from his time of service with the Soviet army in Afghanistan were fascinating, his casual racism toward our fellow passengers was less so. When the sun rose on January first, 1995, I made a solemn promise—which I've kept—to never drink vodka again.

The trip ended with a ride on the branch line leading into Beijing, nine thousand kilometres from Moscow. From there we explored Shanghai, Hong Kong, Hanoi, Bangkok, and points in between, wrapping up our itinerary on the beautiful Thai island of Ko Samui, where my father had taken my brothers and me a few years earlier. To commemorate the journey, I had a local artist tattoo an image of the earth on my left shoulder.

Finally, I made my way back home late that spring, first to Vancouver to visit my mother's family, and then to Whistler, where Michel was living and working at the time. My return to Canada gave me much cause to reflect on my year away from the only country I could ever call my own.

No journey so extensive and all-encompassing can leave a traveller unchanged, and I was no exception. Like most Canadians who have been fortunate enough to travel abroad, I came back with a heightened appreciation for our country's unique mix of blessings. I couldn't articulate everything I experienced in the trip and catalogue all the ways it had influenced my point of view. The change was general and broadly based. It deepened my sense of our need for awareness and understanding of people from different backgrounds, and my conviction that if we choose to emphasize it, the common ground we share can dwarf any difference. I had also had plenty of opportunities to observe that communities where people are open to difference, to others, are happier and more dynamic than places that are more insular and closed off.

The incredible diversity I experienced while steadily travelling east for a year caused me to notice something that I had taken for granted at home. Wherever I went, there were locals. A clear majority. A mainstream. And any minorities, be they North Africans in Paris, European expats in Burkina Faso, Lebanese supermarket owners in Ivory Coast, Chinese students in Russia, Australians in Thailand, or even tribal or cultural minorities that made up a significant chunk of the country's population, were always "others," an exception to the rule, to the national identity.

In contrast, our modern Canadian identity is no longer based on ethnic, religious, historical, or geographic grounds. Canadians are of every possible colour, culture, and creed, and continue to celebrate and revel in our diversity. We have created instead a national identity that is based on shared values such as openness, respect, compassion, justice, equality, and opportunity. And while many of the almost one hundred countries I've travelled through in my life aspire to those values, Canada is pretty much the only place that defines itself through them. Which is why we're the only place on earth that is strong not in spite of our differences but because of them.

THAT SUMMER AND FALL IN MONTREAL WITH MY DAD, I got to know my four-year-old half-sister, Sarah, a bit better. Her mother, Deborah Coyne, was a constitutional lawyer who was a good friend of my father's.

I had seen Sarah a few times when she was a baby, and was glad to see her a few more times as a precocious little girl. Truthfully, it was a delight to see my father, approaching eighty, carrying Sarah around on his shoulders as he had done with my brothers and me when we were her age.

In September 2000, a few days before Dad passed away and after Sarah and Deborah's final visit with him, I would take Sarah out rock climbing—an activity that I know my dad would have loved to have seen us do together.

After the funeral, with Dad gone, we lost touch. I remain proud of my half-sister and look forward to connecting again in the future.

MY FIRST YEAR BACK AT MCGILL WENT WELL, WITH new courses and new friends, but in my second year I became unmotivated. I loved the classes and the teaching experiences, but after some honest reflection, I realized my life was in a bit of a rut. I was still living at home with my father, and as much as I loved him, I needed to move out on my own. Which, I rapidly concluded, also meant leaving Montreal.

My trip had had a deeper effect on me than I'd thought.

When you travel away from a place where you have spent many years of your life, you leave behind a negative space, an empty contour of the person who left. When you return from your travels, you expect—and are expected—to occupy that same space again, but it never quite fits you, because you've changed. It's not only uncomfortable for you but mildly disconcerting to those who know you well.

I had, I knew, returned from that journey different from the guy who left. Now Christmas 1996 was approaching. It would mark my twenty-fifth birthday, as good a time as any to step outside the comfortable space that my friends, my family, and my personal experiences had created for me in Montreal.

Never less than realistic and honest about life, my father understood and agreed with me. Sacha was still living with him in the house on Avenue des Pins, so Dad wouldn't lack for company when I left. But where to go?

The answer came easily.

Through all the family trips to British Columbia we had taken over the years, I had dreamed of living on the West Coast. The scale of the West had been both intriguing and a little intimidating to me as a young boy, exemplified by the coast and the mountains and by those massive trees—the giant Douglas firs of Stanley Park that the three young Trudeau boys, arms outstretched and connected at fingertips, could not even half encircle. But what really drew me out west was family. My Sinclair roots, and my brother Michel, who was by then living in the Interior. In January

1997 I headed for Vancouver on my way up to Whistler. My plan was to find work as a snowboarding instructor.

Skiing was in our family's blood. Every one of us was a solid skier, and each had an individual style. My father's was strong, aggressive, and very clean. My mother had learned to ski at Whistler as a child, and she had beautiful form. She prided herself on never falling, and I cannot recall ever seeing her take a spill. Sacha, Michel, and I all learned how to ski when we were barely out of our toddler years. Of the three of us, I have to confess that Michel was the best, perhaps because, while trying hard to keep up with his older brothers, he developed skills that we didn't need. Sacha hewed closest to our dad's tracks in all things, including skiing, and developed the most elegant technique. My approach was more basic. I never quite got the aesthetic rhythm of turning. My goal was always to get down to the bottom of the hill as fast as possible, which generated a steady stream of spectacular wipeouts.

When it became obvious that I would never distinguish myself within the Trudeau family with my skiing ability, I decided to pursue something new. When I was fourteen, I was inspired by the opening sequence of *A View to a Kill*, in which James Bond rips a ski from the front of a snowmobile and rides it as a snowboard. I promptly mail-ordered a board from Vermont and taught myself how to use it at Mont Tremblant. And so, when relocating to British Columbia in my twenties, I intended to resume snowboarding in the interests of both fun and profit.

Before becoming an instructor I would have to obtain level one certification in the sport, which would take some time. To cover my room and board (not so much room as a mattress in a friend's loft, and not so much board as a few daily slices of Misty Mountain Pizza), I managed to get myself hired as a doorman at a popular nightclub called the Rogue Wolf. I enjoyed the work, so I kept the nightclub job even after I earned my certificate and began working at the snowboarding school. My schedule was practically non-stop. For six days a week I was at the snowboarding school, starting early in the morning and working until five. And four nights a week, after a few hours of rest, I did a shift at the Rogue Wolf, where I usually worked until two or three in the morning.

I loved the schedule. Responsible for kids all day, responsible for keeping the peace all night. And at no point did my last name come into play.

Of all the guys working the door at the Rogue Wolf on the busiest nights, I was the smallest. One of the others, Peter Roberts, who remains a good buddy to this day, had been with the Canadian Forces and had even trained my brother Sacha at CFB Gagetown when he was in the reserves. Pete imposed respect easily through his mere presence, but I had to come up with other ways of getting things done. Despite being less physically intimidating than the others, I was usually designated the first responder when situations at the club began turning sour. If a biker walked in without paying the five-dollar cover charge, I'd be sent to collect the cash. At first I suspect they were just hazing the rookie, but later I'd get sent in first because I usually got results without confrontation. Needless to say, during my time at the Rogue Wolf I learned a lot about human nature.

I discovered that the secret of an effective bouncer is to be diplomatic and unintimidatable. It's also important, of course, to be sober in both the literal and abstract senses of the word. By keeping his wits about him, a good bouncer can almost always avoid having to engage in any sort of physical skirmish. In my case, my wits were my most important asset. The gigantic bouncers you see outside some popular night-clubs don't care very much if things become physical; they can just bear-hug unwanted patrons, carry them through the club, and dump them outside. I couldn't do that, and I wanted to avoid punches. Any time a punch was thrown my way, it meant I had screwed up by failing to resolve the situation firmly but peaceably.

My favourite trick when dealing with an intoxicated would-be brawler was to say, "Look, pal, you don't want to take me on in here because these other bouncers will jump on you. So if you come outside with me, we'll settle this between us."

Eager for a fight, or at least for the *appearance* of a fight, the guy would immediately oblige and head for the parking lot, on the way bragging to everyone who cared to listen how he was going to whip my butt or clean my clock or some other description of the beating I could expect. I never gave him the chance. I would see him out the door, hand him his jacket, flash him a smile, and wish him good night before returning inside, which usually incited him to swear more loudly and call me a chicken.

"You're right," I would call to him. "I don't want to fight. Now go home and sleep it off, and I'll see you tomorrow night."

The lessons learned at the Rogue Wolf were broad enough to have some practical applications in politics. Whether you are trying to assert your will in a barroom confrontation or a political altercation, the biggest obstacle to overcome is the human ego. Once a disagreement begins, no one wants to back down. The trick is to find a way for your opponent to save face, like leaving the aggressive drunk waving his fist in triumph, but in the rain. Meanwhile you're inside, staying warm and dry and getting your job done.

Along with the chance to develop and practise some basic psychological tactics, working at the Rogue Wolf gave me a glimpse into the ways that young people can self-destruct through their use of alcohol and drugs. I watched too many people do too many stupid things simply because they were bored, and I saw too many testosterone-fuelled young men assume that the only way a night out can end successfully is in a fight. Being dependent on drugs and alcohol for your happiness is a trap that has ruined too many lives, and I resolved long ago that it wouldn't ruin mine.

There were equally valuable tactics to be learned from my other job, teaching kids how to handle a snowboard. We instructors functioned as part of Blackcomb's innovative "Ride Tribe" teen-training program, which hadn't proved very popular in the beginning. The Ride Tribe was considered a holding bin for kids who didn't have any snowboarding friends or were too old to be happy riding while their parents skied. My friend Sean Smillie changed this image with a program he developed from scratch. He started by choosing a handful of instructors who he knew loved to teach kids and found new ways for them to pass their knowledge on to their students. The success of Sean's program taught me how innovative teaching methods and a high-energy teaching staff can motivate even the most jaded students.

Sean recognized that snowboard instruction didn't have to follow the old ski-school model of turn-stop-turn-stop-repeat. Snowboarding is an exciting sport, maybe the most exciting activity on snow. In the right milieu and with the right instruction, it offers a steeper learning curve than most sports. After just a week on the mountain, most beginning snowboarders are able to perform carving manoeuvres and simple tricks, which is why the sport grew so rapidly in the late nineties. Sean capitalized on this by recruiting teachers whose aggressive snowboarding techniques challenged their students.

And it worked. Kids who passed through our program began to tell their friends about it, and soon we heard of copycat programs starting up at places like Vale and Aspen. The Ride Tribe program had always been a barely break-even proposition at Blackcomb. Now, almost overnight, it turned into a money-maker. The magazine *Teen People* sent folks to do a feature on Ride Tribe, giving it the sort of promotional bonanza that my Ottawa colleagues in politics might refer to as "earned media."

Any teacher will tell you that the most rewarding moments of their profession occur when the light bulb switches on—when they witness a student suddenly "getting it," whatever *it* happens to be. In snowboarding, those light-bulb moments arrived several times a day for me, and each time they gave me a kick. Part of the reason for these sudden insights lies in the nature of snowboarding compared to skiing. Skiing instructors essentially communicate tips to their students, but teaching how to handle a snowboard involves revealing *secrets*. Tips are fine, but secrets are sensational. When the kids managed to turn a snowboarding secret into a new move, they could barely contain their excitement. For the teacher, it was the equivalent of watching an entire class suddenly understand trigonometry just by rotating their hips.

One of the biggest challenges I faced as a snowboarding instructor was to make my know-it-all teenagers aware of all that they *didn't* know about the sport and about the alpine world generally. This illustrated another important distinction between skiing and snowboarding.

Skiing takes years to master, and the time gives everybody on skis a chance to become familiar with the risks and rhythms of traffic on a busy hill. The same isn't true of snowboarders, some of whom might be rocketing down intermediate and advanced runs after only a few days. They don't yet know the etiquette required when merging trails or the safest places to stop on a steep hill. This is one reason skiers often complain about the bad manners of snowboarders. It's not that boarders are inherently boorish; it's just that they usually have less experience in alpine environments (and the blind spot created by travelling sideways doesn't help, either). Each time I rode up on the chairlifts with students, I would ask them to look down and predict which skier or snowboarder on the hill would go where, who would stop to rest, and who appeared most likely to crash. It was akin to teaching defensive driving; I wanted my students to be aware of everything and everyone around them.

My experience at Blackcomb gave me insights into the art of controlling large groups of children. Parents started dropping their children with us before eight in the morning, but we wouldn't head up the mountain until nine, leaving us an hour to spend with a gaggle of kids aged twelve to sixteen. As anyone who has been around teenagers knows, kids at that age tend to roll their eyes at authority figures before either wandering off or making trouble. So while I organized loads of activities for them, I knew my most important role was to project confidence and leadership. If I let it waver, even for a moment, there was a good chance I'd lose the group before we took a single snowboarding run. Before long, I realized I had a talent for engaging these kids.

The biggest effect my experience as a snowboard instructor had on me was to point me solidly back toward teaching. All the joy, all the satisfaction, all the fulfillment I felt at the end of each day with the snowboarding class convinced me that I had much to offer as a teacher and that teaching, in turn, had much to offer me. I had been a camp counsellor, a whitewater river guide, a snowboarding instructor, a bartender, and a bouncer. All these positions had left me wondering if I would ever be happy with a "real" job. Teaching was very much a real job, and I was eager to get started once again.

When I discussed my rekindled interest in teaching with my aunt Heather, she informed me that I could qualify for the University of British Columbia's twelve-month education program. At that moment, everything became clear. At the end of the ski season I headed back to Montreal to pick up a few prerequisite courses at McGill, then said my goodbyes to friends and family, and returned to Vancouver with a new focus.

I FORMED MANY NEW FRIENDSHIPS IN UBC's FACULTY of Education. The students were a spirited and diverse bunch, and with much of the instruction based on interactive classroom training, we had plenty of opportunity to learn from each other.

That first year in Vancouver passed quickly. During the week, I worked hard toward securing my education degree in the city, then escaped to Whistler on weekends. Almost before I knew it I had graduated and was substitute teaching in Coquitlam. By then I had moved into a larger apartment with my UBC classmate Chris Ingvaldson and his wife. Chris introduced me to the headmaster of West Point Grey Academy, and soon after I was offered a full-time teaching position there. (Chris would also become a teacher at the school.) West Point Grey, a private co-ed school founded just two years earlier, was a good posting for a newly minted teacher. The staff was young and energetic, and it was great to be a part of building a new school's culture.

Over ten years later, long after I moved back east and entered political life, I got a late-night phone call with shocking news: Chris had been arrested on a charge of possession of child pornography. Eventually he pleaded guilty and was sentenced to three months in jail. Like everyone else who knew Chris at teachers' college and in subsequent teaching jobs, I was utterly shocked. Chris lost his teaching job, his marriage, and most of his friendships, including mine.

Whenever the media announce that someone is facing charges involving child pornography or similar offences, they often feature interviews with neighbours and colleagues who say that the defendant gave no clue about his sexual predilections. Like many people, I had always been somewhat skeptical of those testimonials. I assumed that you should somehow *know* when you were in the company of someone with such tendencies. After the episode with Chris, I realized that isn't true. It also explained why police officers and prosecutors need to work so hard to protect our children from this kind of exploitation. Neither Chris's wife nor I as his roommate ever envisaged the dark path he would lead himself down. It was a bitter lesson for me, and one that needs sharing.

I SPENT TWO AND A HALF YEARS AT WEST POINT GREY Academy, teaching mostly French and math, but from time to time other subjects like drama, creative writing, and a Grade 12 law class.

In each of my classes, I tried to avoid that "sage on the stage" method I had experienced at Brébeuf by inserting collaborative intellectual exercises into my lessons. These included math puzzles and brainteasers, which have always been something of an obsession for me. Here's an example I remember discovering during my early days in Vancouver: the well-known (among mathematicians) "7-Eleven problem." It's based on a customer entering a convenience store, choosing four items, and watching as the clerk totals the amount owed on a pocket calculator. When the clerk announces that the amount is $7.11, the customer points out that the clerk hadn't added the prices but had multiplied them. Apologizing, the clerk is careful to add the prices this time, hits the Total key—and is amazed to discover that the total is still $7.11. The challenge is to determine what the exact prices of those four items must be to produce $7.11 when both added and multiplied. Solving the puzzle isn't child's play, and if math puzzles are not your thing you probably couldn't care less. But if this kind of brainteaser appeals to you, you could find yourself devoting days to finding its solution. As I did.

Here's my point: if I could find ways to engage my students in tackling their lessons the way that the 7-Eleven puzzle engaged me, it would surely generate the light-bulb responses that I had loved to see during the snowboarding lessons.

Some of the techniques I used were simple yet effective. For example, when teaching math I would start a class by asking why our numbering system is based on 10. Why not a base of 8, for example? Or a base 6 or some other number? Virtually every culture in the world has based their numbering system on 10. "Why is that?" I would say, then ask everyone to raise their hands and look around. One by one the students would realize that their outstretched hands with their ten fingers weren't signalling that they knew the answer—they *were* the answer.

Puzzles worked especially well in algebra. I would tell my class that a father and his daughter go fishing, and when they are about to return home, the father asks his daughter to give him one of the fish she caught. "Then we'll have the same number," he explains. The daughter responds that if her father gave *her* a fish, she would have twice as many as he had. So how many fish had each caught?

The solution can be found via two simple algebraic equations with two variables. Of course, there are more intuitive ways of solving it, and the kids who found the solution in the

23. Here I am with Gerry Butts on the steps of McGill University's Arts Building, a great spot to hang out with friends overlooking the campus and city. We had met on similar steps outside of the student centre a few years earlier.

24. This overland truck was home for our trek across Africa in 1994—no better way to see the continent.

25. Grinning on the night of the 1995 referendum after I tried putting my arm around the police officer and was firmly pushed back.

26. With students at West Point Grey Academy in Vancouver, where I was a teacher in the late 1990s and early 2000s.

27. Kayaking with Miche, on the same route on the Rouge River I took with my dad, pictured earlier.

28. His dog Makwa in tow, Miche spent the summer after his car accident in 1998 reconnecting and rebuilding all the relationships with his family. We later understood it had been a chance to say goodbye, because he died that fall.

29. This was the only time my father managed to visit Kokanee Lake. It was the September after we lost Michel, and he marvelled at the great beauty that now surrounds his youngest son.

Sophie and I love to explore and to stay active. As much as she looks out for me politically, she can also be quite the competitor, no matter what the sport.

30.

33. I am very lucky to have the friends that I do, and even luckier to have them welcome Sophie with such open arms. Pictured here, left to right, are Ian Rae, Gregory Ohayon, Marc Miller, Mathieu Walker, Tom Pitfield, Gerry Butts, me, Seamus O'Regan, Allen Steverman, and Kyle Kemper. Also part of the wedding party, but not captured in this shot, were Navid Legendre and Sacha.

34. I was welcomed with equally open arms by Sophie's wonderful family.

35. The nomination contest in Papineau in 2007 was my first major political rally as a politician rather than an observer. I always feel lucky to have Sophie by my side, but I felt even more so here because she was pregnant with Xavier at the time.

36. The family together at Sacha and Zoë's wedding in 2007. Pictured here, left to right, are Sophie, Alicia, my mother, me, Zoë, Sacha, and Kyle.

37. Here we are bringing Xavier home for the first time. This photo captures that point in all new parents' lives when they realize they are responsible for a little being and no longer have the help of nurses at the hospital—it is stressful but wonderful.

38. Ella-Grace came along less than two years after Xavier, bringing even more joy to our family.

39. We don't bring the kids out to a lot of political functions, but when we do, we always try to strike the balance of spending time with them while we work.

40. Family time is essential to what I do. It grounds me and reminds me of what we are working toward when on the road. Here Sacha and I gather with our families, along with Sophie's parents, Estelle Blais-Grégoire and Jean Grégoire.

41. Mom happily showing the kids a magazine we were in. They have always been fascinated with the idea that the family shows up in newspapers and magazines from time to time.

42. My dad used to do similar stunts with us, and I guess he passed that along. The kids don't seem to mind either.

shortest time weren't always the math whizzes. The point was to engage them, to lead them to the "Aha!" moment when the light bulb came on and made their faces glow.

I tried to take the same approach when teaching English. To show the students the rhythms of poetry, I would write "toobie hornet toobie" on the blackboard and see who could flip the accented syllables to declaim a line from *Hamlet*. There was a little bit of my old history teacher André Champagne in my approach. Like him, I wanted students to think about things they took for granted, such as the way we count and the way we pronounce simple words. My goal was not just to teach students a certain body of material but to also give them the critical-thinking skills they would need to work problems out on their own throughout their lives.

My often unorthodox approach puzzled school administrators a little, but it made me popular with many of the students. They knew I was a sucker for a good debate, and they would sometimes lure me away from drilling the core curriculum into their minds by engaging me in some tangential philosophical discussion. This led to the odd difficulty. When I needed to apply discipline to my students, they would sometimes interpret it as some sort of a betrayal on my part, thinking that I had turned on them.

My response was always, "I like and respect you, which is why I have high expectations of you. If you don't do the homework or if you flunk a test, there are going to be consequences. You need to know this. You need to be aware of the expectations you'll have to meet in life. I wouldn't be much good if I didn't help you deal with these realities by holding you to account for what you do."

WHENEVER I SPEAK TO PEOPLE WHO DON'T WORK IN THE education system, many assume that it is easier to be a teacher in a well-funded private school than in the public system. There is some truth to that. Private schools may have newer and better amenities, and discipline tends to be more effective. But there are trade-offs, especially when it comes to working with parents to get the best out of their children. After paying thousands or even tens of thousands of dollars in tuition, the parents of private school students may become upset if they fail to see results (that is, high grades). And teachers may censor themselves in parent-teacher interviews, concerned that an angry parent may choose to complain to the administration or threaten to send his or her child to another school. I always preferred to speak the truth to parents. If I thought a child's school performance was suffering because of his or her home environment, I would say so. Which now and then ruffled some feathers.

Sometimes my teaching methods put me at odds with the conservative West Point Grey administrators. The most significant event concerned a student I'll call Wayne, who regularly defied the school's dress code by wearing his tie loose and dangling a chain from his belt. He was a smart, confident kid who chose to adopt a rebel pose. One day, after receiving the umpteenth dressing-down about his attire, Wayne said to me, "It's not fair. I'm always being called out for my appearance, but the same rules say that the girls' kilts are supposed to be no more than an inch above their knee. They flout that rule and get away with it. It's a double standard."

I was in charge of the in-house student newspaper, a task I undertook with the clear intention to turn it into something kids actually wanted to read, not just a glossy feel-good pamphlet for proud parents. I suggested to Wayne that he write an article on the unfair double standard he had grumbled about. He did, and his article reasonably theorized that perhaps the predominantly male teachers felt uncomfortable pointing out to teenaged female students that their skirts were way too short. It was the sort of thing that everyone knew but no one admitted, until Wayne mentioned it.

When Wayne's article appeared in the school paper, the administration didn't react as well as they could have, in my opinion. They not only disciplined Wayne for lacking respect but also discontinued the student newspaper, which convinced me that West Point Grey was not the best fit for me as a teacher, nor I for them. Shortly after, I took a teaching position in the Vancouver public school system.

I want to emphasize, however, that overall I loved my time at West Point Grey. The students and teachers were bright and engaging, and the great things about the school vastly outnumbered its challenges, even in the early years. It also makes me really happy to see that by all accounts West Point Grey Academy continues to stand out as one of the top schools in B.C.

My experience had revealed negative aspects of private education that I had not observed during my own schooling. At Brébeuf, admission had been based largely on an applicant's performance in a standardized admission test. At West Point Grey, as at many Canadian private schools, the biggest barrier to entry wasn't the student's ability but the tuition. While there was some access to scholarships and bursaries, many of the students lived privileged lives, with tennis camps to go to in summer, trust funds to draw upon, trips to Europe to enjoy. None of this on its own made the students difficult to deal with; I met plenty of great kids at West Point Grey. But I worried about the level of materialism being displayed. I would overhear some of them discussing what they would do if they won a lottery, something I found bizarre given that they were hardly children from economically deprived environments. The lottery winnings they dreamed about were always $10 million or $20 million, never $1 million, and the kids would fantasize about spending it on exotic boats and private jets.

Whenever I discuss the problem of income inequality in our society, I think about the children and their families I met when teaching at that school. The parents I encountered at parent-teacher nights were successful, hard-working people, but their wealth gave some of them an excessive sense of entitlement. And many of the students had little exposure to or understanding of the larger society around them and the challenges faced by ordinary people.

I knew many wealthy kids when I was growing up, and I had plenty of advantages myself, including the opportunity to travel with my father. But Dad never spoke of wealth as being the ultimate goal in life. Not once. He would have been appalled to hear me rhapsodize about owning jets and living on Caribbean islands. When he was once asked what values he wanted to pass on to his children, my father replied, "I would want them not to be slaves to material goods. I would want them to appreciate a good meal, a good book, and enjoy holidays. That's fine. But being miserable if one is deprived of material-based pleasures, I consider that a form of slavery." Suffice to say, he was successful in passing along those values to his sons.

The public school system was not entirely free of the kind of thinking that disturbed my father. At Vancouver's Sir Winston Churchill Secondary School, where I taught after West Point Grey, the students tended to come from much less wealthy families, but many of them also seemed caught up with their own materialistic fixations.

At Sir Winston Churchill I overheard one student tell another, "My dad got a new job, so he just bought a Mercedes." He was excited and proud and wanted to share his elation.

"Let's have a chat about this," I said. "A nice new car sure feels great. But at the end of the day, remember that it's just a car, a way to get from point A to point B." Then

I added: "And sometimes a promotion like your father received comes with lifestyle sacrifices. He may have more responsibility, but he may also have more stress too. You might find your dad is going to be a little bit more worried about work. He might need to work harder and for longer hours. You get a nicer car in the driveway, but less time with your dad. There are always trade-offs in life. You just need to think fully about what's truly worth dreaming about."

I have given variations on that speech about quality of life to dozens of kids, and I like to think that I got through to at least a few of them.

THE MOST MEMORABLE AND POWERFUL MOMENT OF MY teaching career occurred on September 11, 2001. I woke at six o'clock that morning to the sound of my roommates knocking on my bedroom door. "Turn on the TV!" they called to me, and I tuned in just as the second plane struck the towers. I dressed and went to school, knowing that every student in class that day would have seen the same images I had just witnessed.

"Obviously, we're not going to be talking about French grammar today," I said to my Grade 9 and 10 students. "Let's talk about what happened a few hours ago."

One student asked, "Does this mean World War Three?" Since the leading theory was that the United States had been attacked by Islamist terrorists, this was a logical question. Other students said they didn't think the events in New York had much to do with Canada, and one asked, "How will this really change *our* lives?" just as an aircraft flew low overhead in a way that we had never heard before. It was obviously a military plane, and the class became ominously silent. "That fear instinct you just experienced, that's what's new," I said. We talked about terrorism and the need to fight it, along with the need to ensure that vigilance didn't become a form of paranoia directed at all Muslims. The kids were all shaken by the events, and as a teaching staff we did our best to both address it and draw them back into the routine of everyday life. It was a challenging time.

Later that day, I got a call from Gerry Butts, who had been in California with his wife, Jodi, and was trying to return to Canada. American airspace was closed to civilian air traffic, so Gerry and Jodi rented a car to drive north. They had to drop it off at the town closest to the border and cab it to the crossing, where I would pick them up.

When I arrived at the Peace Arch, I encountered an absolute mob scene, with cars backed up for hours on the US side of the border. Gerry and Jodi just walked along the highway with their suitcases and astonishingly strolled across the border to my car. After I welcomed them, Gerry said, "Uh, shouldn't we check in with a customs official?" We left the car and wandered around the facility until we found someone able to process pedestrian travellers, which he did in a matter of seconds. As we drove away, I glanced at the long line of cars in the rear-view mirror and thought of all the families with kids in the back seat, waiting for hours or even days, while my friends had been able to cross the border as if it did not exist. The incident symbolized the disorganized and ad hoc nature of the immediate security response to 9/11.

Vancouver is a beautiful place, as far removed from the world's violent hotspots as anywhere on earth. There—as in most of Canada—it's easy to feel both protected and detached. I recognized that in the new era launched by 9/11, no place on earth stood immune from the threats facing our world. Since that day many nations, including Canada, have scored successes against al-Qaeda and like-minded groups. But the threats remain, which is why I urged my students never to forget where they were when they heard the grim news on September 11, 2001. Our memories of that tragedy, painful as they may be, are our best means of ensuring that we remain vigilant in the fight against terrorism.

CHAPTER FOUR

The Woods Are Lovely, Dark, and Deep

———

IN NOVEMBER 1998, I SPENT A WEEK AS A SUBSTItute teacher at Pinetree Secondary School in Coquitlam, about a half-hour's drive east of Vancouver. The class had been a good group of kids, and by week's end I was sorry to leave them. After saying my goodbyes on Friday the thirteenth, I drove back to my apartment, had dinner, and went to bed. I fell to sleep unaware that earlier that day, I had lost my little brother Michel.

My telephone rang at five the next morning. It was my mother calling to say there had been an accident, and I knew from the tone of her voice that it involved one of my brothers.

I discovered that the cliché phrases were all based in reality: I went numb, my heart sank, and my blood ran cold all at the same time. "We don't know for sure, because they still haven't found him," my mother said. "But the RCMP just told us that Michel was caught in an avalanche up at Kokanee."

Michel had been doing what he loved most when he died, backcountry skiing with friends in the Southern Interior of British Columbia. While I had been standing at a blackboard, an avalanche had swept my brother and one of his buddies into Kokanee Lake. They had been traversing the steep incline above the lake. His pal Andy managed to swim to shore, but Michel was just too far out. It had taken his other friends hours to dig themselves out and contact the RCMP. Meanwhile, I had had a normal day, as had the rest of my family back east, as yet blissfully unaware of what had happened.

Part of me was certain that Michel was still alive. I just couldn't conceive of a world in which he wasn't.

I felt a spasm of guilt. What was Michel *doing* out on that glacier? Why hadn't I, as his older brother, found some way to protect him? We lived in the same province. I should have visited him more, called him more, watched over him more, done *something* to keep him from danger.

It had already been a trying year for Miche. While he was driving home through Manitoba in the spring, a careless driver caused an accident that totalled his truck. He escaped serious injury and was most concerned about his dog, Makwa, who had run off from the accident; it took a week to find him. Compounding his hassles, when the police arrived at the accident scene they found some marijuana in the glove compartment, for which he was charged. But perhaps because of that near-death experience, Michel devoted much of the summer to reconnecting with the family, rebuilding the loving relationships that we had all let drift because of geography and busyness.

When fall came, we three brothers again went our separate ways, and that was the last time I saw him, although he wasn't too far from me, working on the ski hill in Rossland.

I had spoken to him on Monday of the week he died, a telephone call I made partly out of guilt. I had been kicking myself for not calling on his birthday in early October when a friend reminded me that, when it comes to family, it's never too late to connect. She was right, of course, and I called him later that very day. We had a good talk about many things, the usual back-and-forth chatter between siblings. The subject that I remember most clearly was his plan to spend three days later in the week up on Kokanee Glacier.

"It's early in the season," he said, "so we have to be careful."

I replied in the assertive tone of a concerned parent or older sibling: "Yes, you must be *especially* careful at this point in the season." He burst out laughing. He knew that I knew little about avalanche dangers and the steps that need to be taken to avoid them. I knew only that mountain skiing in that region of B.C. always brought the risk of avalanches. I learned so much more when, after Michel's death, I became a director of the Canadian Avalanche Foundation and pushed for increased funding to support avalanche awareness.

When word of Michel's disappearance spread later that day, the national news media flooded Rossland, eager to get a quote from anyone who knew Michel. All the comments were the same—he was a happy-go-lucky young guy they knew only as Mike, popular with everyone who met him, a bon vivant type with a quick smile. They were all surprised to learn that this well-liked fellow who had no pretensions and intensely loved exploring the wilderness on skis was the son of a former prime minister.

Michel had built so much of his life around the snow and the air and the mountains and the people around him. He was a free spirit, enamoured of the Aboriginal culture that flourishes in Canada's most beautiful places, a guy utterly at peace with himself in a way that had so far eluded both Sacha and me.

After booking an early flight to Montreal, I called my father to let him know I was coming and to ask if he had any news. My father had never been given to hopeful self-delusion, and he wasn't now. "No," he told me sadly, "and there won't be any news because Michel is gone. The only question now is whether or not they find the body."

Michel had ventured to Kokanee because the area held all the attractions he valued in life: a remote wilderness location, stunning scenery, challenging skiing, and the kind of stillness that is so rare in our hectic world. Skiing on the glacier above the shore of Kokanee Lake on a perfect sunny day was close to paradise for Miche. Kokanee Lake itself is an alpine jewel about a kilometre long, four hundred metres wide and very deep, surrounded by cliffs and precipitous rock slides. I understand why Michel would be drawn to it and how he probably weighed the risk of an early-season thaw, despite his laughter at my concern. The danger was not really an impediment. If he wanted badly enough to challenge his skill and satisfy his need for adventure, he would have gone under almost any circumstances. And he did.

He had probably long before come to terms with the risks faced by adventurers in the rugged parts of Canada where he most felt at home. A few years before his death, Michel had been idly watching a TV documentary about burial rites in Asia when he stated, matter-of-factly, "When it's my turn, just leave me down at the bottom of the mountain where I lie."

Kokanee Lake was at the bottom of the mountain, and the early-season avalanche knocked him off the path and into the depths of the lake. Had it been later in the year, the ice would have been frozen, and he and his buddies would have simply watched from the safety of the track in the centre of the lake as the small slide came down. His comment proved prescient: divers would never find his body, and he is there to this day.

Michel had carved his own route through life. While Sacha and I attended McGill, close to Dad, Michel chose to head east, to Dalhousie University in Halifax, where he studied microbiology. From there he went west for a life in which he wouldn't have to think about the expectations others might have of him.

In my third year at McGill, I drove to Dalhousie to spend some time with Michel. We had been close as kids, but by the time he entered university, we had drifted apart somewhat. In Halifax, he seemed to get away from the influence of his brothers and his father, and he partied hard. It was on that visit that I came to understand his desire to forge his own identity.

I still miss him. I will always miss him. Michel was just twenty-three when he died, but he had already found his calm zone, a private place that eludes many of us throughout our lives. If Michel were alive today, I believe he would be the father of teenage children, and that Sophie and I and our kids would visit him and his family each Christmas. Perhaps Michel would have launched his own ski tourism operation; he loved the sport, and I believe he had a knack for business. In his spare time, he would have found a way to bring out his creativity through painting or writing. Whatever he chose to do, I know he would have been happy doing it. That was his gift.

WHEN I STEPPED THROUGH THE DOOR OF MY DAD'S house in Montreal late in the afternoon of the day we heard Michel was lost, I hugged my father tightly. Before we could speak more than a few words to each other, the telephone began to ring. My father made a move to answer it, but I stopped him and said, "No. That's why I'm here."

Once released, the news of Michel's death sped across the country almost instantaneously. While my father dealt with his private grief, I spent the rest of the evening accepting condolences from family friends. The calls continued for several days. They were sincere and touching, and in a way they helped me deal with my severe emotional pain. I was fulfilling a function by helping my father through the worst days of his life, and the duties I assumed helped dilute the agony I was feeling over my brother's death. Sacha had been filming a documentary in the Arctic when he got the news. He flew to B.C., where he acted as the family's representative amid the rangers and divers involved in the recovery effort at Kokanee Lake, and he made a point of thanking each one of the searchers on our behalf. My mom was home in Ottawa devastated, comforting and comforted by Fried, Kyle, and Ally.

I stayed with my father in Montreal, helping to organize the memorial service for Michel at Église Saint-Viateur d'Outremont. The activity helped by keeping me busy, for a few days, but as reality set in, emotions overcame me. Sach delivered a beautiful, poignant, but heartbreaking eulogy. I could find no words and so read a First Nations prayer Miche loved.

Afterwards a reception was held at the Mount Royal Club, where some of Michel's friends staged a video tribute to my brother. At first it seemed incongruous to have a life as vital and, in its own way, as rebellious as Michel's being celebrated in a well-upholstered club straight out of Montreal's gilded age. But as the rooms filled with Michel's friends, it became clear that the Mount Royal dress code, which leaned heavily on bespoke tweed suits and Italian silk ties, was set aside that day. Michel's friends, from Dal, from Camp Ahmek, from out west, arrived dressed as they were, and no one complained. The ambiance was sad, and beautiful, and the smell of patchouli was everywhere. It warmed my broken heart to be surrounded by people who had loved Michel so deeply, and so well.

Over the weeks that followed, Sacha became my father's principal pillar of support after I moved back to Vancouver to resume my teaching duties and my life, trying to put the tragedy behind me. But at Christmas more bad news arrived, when my father was hospitalized with a serious case of pneumonia.

That was the medical diagnosis, at least. My own opinion was that the lights began to dim in my father's soul when Michel died. He recovered from his pneumonia within a few weeks, and even travelled a bit after that. But from the time we buried Michel until his own passing two years later, my father was never the same man.

My mother endured horrific, debilitating grief at losing her son, compounded by and compounding her mental health issues. She went through an extremely difficult period that left her entire family struggling to help during the five or six years that followed his death.

Michel's death and its impact on my parents affected me deeply. I spent long days in contemplation and long nights dealing with the loss of Michel, my mother's struggles, and the reality of my heretofore invincible father's deterioration. I sought and received help from many quarters: faith, therapy, and, most of all, an incredible circle of friends. It was then that I understood that friendship is not truly about being there for the good times, the fun, the adventures, but about being there for each other during our most difficult, solitary moments. It was through that dark time that I understood that the extraordinary people I get to call my friends are what make me the luckiest person in the world.

IN HIS LAST YEAR OF LIFE, MY FATHER GREW MELAN-choly and existential. He had wrapped his mind around timeless questions of human mortality and the fate of the human soul. Sometimes he seemed angry with God, unable to understand why God had taken his son, who was so full of life, instead of him. It weakened his faith. One day Dad suggested to my mother, "If there is no afterlife, then nothing I have done in my life matters at all." It may have been the most profoundly sad thing my father ever said.

During this period, I began to examine my own relationship with God. My father had been a devout Catholic throughout his life. When we were children, he took us to church on Sunday as often as he could, and as a young adult I performed the rites of worship just as they were taught to me. Yet as I grew, all too often the rituals struck me as more ceremony than substance. Was I too childish to appreciate their significance? Perhaps I was. Children dressed in their Sunday best at church are, after all, still children. When Sacha, Michel, and I had attended church services with our father, we dealt with the boredom by trying to make one another laugh without laughing ourselves.

When I was eighteen, I had a long conversation with my father about my attitude toward religion. I told him that I believed, as I believe now, in the existence of God and in the values and principles universal to all major religions. It was the dogmas of Catholicism that I struggled with, particularly the idea that someone who was not a sincere and practising Catholic could not gain entrance to heaven. That seemed strange and unacceptable to me. My father's response to my questioning was to say, "You must make your own choices," suggesting he was content that I at least had a grounding in Christianity, and that I could return to the teachings later in life if ever I so chose.

Michel's death made my father question his faith, but it had the opposite effect on me. Amidst all the searing emotional pain I was feeling, I had a moment of revelation: despite all the torment and confusion we suffer in this *valle lacrimarum*, a divine sense of the universe exists, one we cannot comprehend. With this revelation came an oddly empowering sense that my life, like everyone else's, is in God's hands. This awareness hasn't absolved me of the need to struggle for a better world and a better self, but it has helped me deal with things I cannot change, including death. It also helped reaffirm the core of the Christian beliefs I retain to this day.

Going through this spiritual crisis in the wake of Michel's death, I became friends with Mariam Matossian, an Armenian Canadian who was a teacher at the time but later would become a successful folksinger. Mariam and I developed a genuine friendship, talking regularly, mostly about matters of faith. I was a lapsed Catholic and she was an evangelical Christian with doubts, and both of us were enduring a period of personal reflection.

When Mariam invited me to accompany her to an Alpha course, a program of instruction that guides attendees though discussions of the meaning of life experienced from a Christian perspective, I hesitated. I was suspicious that the course would consist of proselytizing for one sect or another. But I discovered this wasn't the case at all. Instead, it was about developing the humility necessary to admit that we cannot get through life's most difficult challenges on our own. Sometimes we need God's help. I understood that I was going through just such a period, and the course helped me welcome God's presence into my life.

THE SYDNEY OLYMPICS WERE ON TV IN SEPTEMBER OF 2000 when my father died, and to this day, the mere memory of seeing the Canadian flag at half-mast in the Olympic

Athletes' Village the day after still causes me to blink back tears. IOC vice-president Dick Pound said on air that his friend Pierre Trudeau "hadn't been old long," a phrase that perfectly summed up the end of my father's life. Dad had remained a great outdoorsman well into his mature years, able to overcome almost any obstacle he faced. In his seventies he ripped his knee by stepping into a hole while on a vacation in the Caribbean; within a few years of the surgery, he was back skiing the black diamond trails at Whistler. One of the jokes among our family was that whenever Dad went to the movies he insisted on getting his senior citizen's discount. It was laughable to view him as a traditional senior citizen; he was one of the most robust people I ever knew. Until, very suddenly, he wasn't.

A few years earlier, just before I turned twenty-five, at the urging of a dear friend I had had "the talk" with my father. He was still hale and hearty, but as the eldest son I felt that I needed to have a conversation about end-of-life issues, preferably well before they arose. I asked him what sort of care he wanted, and what level of intervention he desired if his body started to fail him. He talked about joining his parents and grandparents in the family tomb in the small town of Saint-Rémi, and said that he wasn't bothered by the likely very public, possibly state funeral he'd have, as long as it was in Montreal.

It was a difficult conversation for me to have with him, but he seemed somewhat bemused by it. I suppose it is a difficult conversation for anyone to have with an aging parent, but on the whole I felt fortunate that, unlike some adult children wrestling with their parents' mortality, I had not come to him seeking some kind of emotional closure. It was a very practical conversation, extremely matter-of-fact, and it made a world of difference to have had that chat before he fell ill.

Of all the memories I have of my father and of our relationship, none is warmer and more poignant than what happened a year before he died, when he came to visit me while I was teaching at West Point Grey Academy in Vancouver. It was a quiet Friday lunchtime, and he enjoyed meeting my teaching colleagues and touring the school with me. It felt good to show him my home classroom and share what I was doing with my professional life.

As we were about to leave the building, we heard the scurry of running feet approaching from behind. We both turned to see one of my students, almost out of breath from chasing after us. As she approached, suddenly nervous, she said, "Mr. Trudeau . . ."

I had seen this sort of scene unfold thousands of times. Everywhere I had gone with my father, star-struck children and adults alike approached him to seek his autograph or shake his hand or ask if he would pose for a picture with them. I would always stand back smiling silently while my father politely indulged the request, and I stood back now.

But this young woman, possibly born the very year my father had taken his famous walk in the snow, didn't even glance at him. Instead she addressed me. "Mr. Trudeau, I just wanted to let you know that I'll be late for French class this afternoon because I have to help out in the gym." I nodded and thanked her; she turned and trotted away without another word.

I felt a little embarrassed by the encounter. This student was the child of immigrants, part of the wave of newcomers who had come to this country and made a success of themselves thanks in part to the open-minded policies my father had introduced as prime minister. Now, he had been treated like some anonymous bystander, and I cringed a little before turning to Dad, unsure what to say.

To my delight, he was wearing a broad smile. After many years of receiving recognition and gratitude for so much that he had done, he hardly needed one more gesture of acknowledgement from a young Canadian. Instead, he had taken fatherly pride in seeing his son maintaining our family's legacy of service to Canada, this time as a teacher of young people. Now *I*, not Pierre Elliott, was "Mr. Trudeau" to a new generation of kids, and he was proud of me for that. It was a lovely warm moment for both of us to share.

And it was one of the last. In the spring of 2000, as I was finishing my school year at West Point Grey Academy, Sacha called to tell me our father was dying. He had been beset with Parkinson's disease and had already survived a bout of pneumonia. He will get past this somehow, I assured myself. But while he was tough, my father was not indestructible. Sacha revealed that Dad had been diagnosed with prostate cancer some time ago and had decided not to pursue treatment. The disease now seemed to be entering its terminal phase.

"What the hell?" I almost shouted through the telephone. "Why didn't you *tell* me?"

Sacha explained that it had been Dad's orders to keep me in the dark. My father knew I would drop everything I was doing in Vancouver and return to Montreal the moment I heard about his condition. He didn't want to me to quit on my students before the school year was over. I know my father had been trying to be considerate, but I was angry anyway. Some irrational part of me thought that perhaps I could have fixed the unfixable if I had known about it earlier. When I settled down, I packed my bags and once again caught a long, sad flight to Montreal, where I would spend the summer with my father, reading him his favourite plays by Shakespeare, Racine, and Corneille, and just sitting quietly with him.

Michel's death had been sudden and shocking. My father's passing happened gradually, week by week, with Sach and me by his side. By the end of September, on a quiet Friday afternoon, it was time, and he let go.

Amidst our grief, I knew we would have to deal with a significant level of media attention as soon as his death was announced. The family home on Avenue des Pins would be surrounded by journalists, as it had been a few weeks earlier when his condition became known. We would not be able to come and go at this private time without the glare of cameras in our faces. Sach chose to stay in the house and hunker down, but I reached for the phone.

I called my old pal Terry DiMonte, a Montreal radio host, and told him I was coming over to crash at his house for the weekend. I liked the idea of hiding from the press in a place where they would never bother me: with one of their own. For the next few days, undisturbed except for the downtown meetings to work through the details of the state funeral with my brother, the government protocol office, and a few of my dad's old confidants, I was able to work through my grief in peace, surrounded by a handful of my closest friends.

It was also over that weekend that I wrote the eulogy. I knew that the newspapers and television tributes would be filled with my father's political accomplishments, so I wanted to share that side of him that people saw but didn't really know: how great he was as a father. My friends helped me recall a few stories that anchored the speech, I put in a few references to the values and vision for Canada that had shaped not just a generation but his own sons, and I tried to offer a little closure to wrap up the outpouring of support from across the country with one last good cry.

The Tuesday morning of the service, as I was getting ready to head to Notre-Dame Basilica, I had Shakespeare on my mind. I thought about the "honourable men" who were my father's political foes, I thought about "praising" versus "burying," and I decided, impulsively, to start the eulogy with a bit of an edge. Perhaps something of a shot across the bow, I can concede now, but at the time, I have to say that I didn't overthink it; it just felt right.

As for the ending, well, I could end only one way: by reminding him, and the world, that I loved him.

And always will.

CANADA LOST PIERRE ELLIOTT TRUDEAU IN THE autumn of 2000. Sacha, Sarah, and I lost our dad. He prepared us well for that eventuality, but you're never really ready to lose a parent. Nobody is. It's one of the biggest changes life presents. Parents are the centre of a person's solar system, even as an adult. My dad had a stronger gravitational pull than most, so his absence was bound to leave a deep and lasting void.

Canadians were overwhelmingly supportive of me and my family. I'll never forget how kind and warm everyone was, almost without exception. Not many people get to lean on more than 30 million others when their dad passes away. At the same time, the change was immediate and overwhelming. When my dad left public life, I was thirteen years old. I went through my teen years and into adulthood in relative anonymity. After my dad's funeral, I was suddenly recognizable to people I passed on the street.

I felt my father's absence acutely. It was sad and profound, but freeing at the same time. I said in my eulogy for him that it was "up to us, all of us" to embody the values he stood for, now that he was no longer with us. Looking back, that advice was for myself as much as for anyone else.

People often ask me if I regret the fact that my father is

no longer around to give me advice, especially now that I have followed in his footsteps as leader of the Liberal Party. Like everyone else who has lost a parent, I miss my dad a lot, but not in this regard. We had a close, deep relationship. For my entire life, he had shared with me his values, his perspective, and his passion, while teaching me to be rational, responsible, and rigorous. Because of that, I feel I need only listen deep within myself to hear his voice in almost any situation.

He's always there in spirit, and his spirit is always encouraging.

CHAPTER FIVE

Two Life-Changing Decisions

———

POLITICS WAS THE LAST THING ON MY MIND after my father's death. I wanted to get back to Vancouver, back to my career as a teacher, and come to grips with the fact that my dad, who had taken up so much space in my life, was no longer around.

In the days after the funeral, I vaguely recall being approached to run for the Liberals, but I made it perfectly clear that I had no interest in doing so. I had a teaching career that I valued, that I was good at, and in which I knew I was making a difference. Entering politics was something that I thought might be a possibility one day, but only if it was on my terms. I had always stayed away from the traditional political world, acutely aware that my name would have far more weight than any of my words or deeds: I was never a Young Liberal, and I never went to conventions or other Liberal events. That world simply did not appeal to me.

I returned to teaching, trading the private system for the public. I threw the weight of my new-found public image behind causes I believed in, such as avalanche safety, but mostly tried to stay under the radar.

My father's old friend Jacques Hébert, who had started Katimavik in the late 1970s as Canada's national youth service program, offered me a position on its board of directors. Frankly, I was a little surprised the program still existed; I remembered the hunger strike Jacques had launched when he was a senator to protest the Mulroney government's axing of it. But I recognized that youth service could fill a void that I had seen in our high schools: opportunities to contribute to and connect with the community at large, giving young people a sense that changing the world in meaningful ways wasn't something they had to grow up first to do but instead could do right now.

In Katimavik, young volunteers worked for not-for-profit organizations and followed an educational curriculum that involved using their second official language, learning about environmental stewardship and Canadian culture, and developing leadership skills. Each year more than a thousand young Canadians, residing in Katimavik houses across the country, contributed to the work of over five hundred partner organizations. Over the life of the program, more

than thirty-five thousand young Canadians have participated in Katimavik initiatives set in more than two thousand
communities. It had an enormous impact on this country,
one that shouldn't be understated. Along with learning the
value of volunteerism and civic engagement, participants
discovered much about Canada by spending the better part
of a year in three different regions with youth from other
parts of the country. Fundamentally, Katimavik was about
young Canadians building a better country, one community
at a time.

My greatest frustration with the program was that
every year, about ten times as many young people applied
to the program as we had funding for. Ten thousand
young Canadians, often unsure about their next step after
high school, would offer to serve their country with their
energy and efforts, and we would turn away nine out of
ten. Any Katimavik alum will tell you it was a life-changing
experience. That a country as successful as Canada would
not choose to offer young people more opportunities to
become active, community-minded citizens while helping
local organizations was something I wanted to fix. And
still do.

After another year or so on the West Coast, I was ready
to head back home to Quebec. I loved Vancouver, had a
great group of friends, loved the mountains and the ocean,
loved the lifestyle, but at thirty, I was starting to feel that it
was time to settle down and possibly start a family. I couldn't
imagine that happening anywhere other than Montreal.

I missed living in French—teaching it just wasn't enough—and it was hard for me to imagine spending the rest of my life with someone who didn't share my language and culture. I also missed, and felt I could help, my mother, who was having an excruciatingly difficult time stabilizing her mental health after the deaths in the family. Living in Vancouver, thousands of kilometres away, I felt frustrated that I was unable to provide my surviving parent with the support she needed.

Finding a teaching job in Montreal was more complicated than I expected. My B.C. teaching credentials would need an overhaul of sorts before I could be accredited in Quebec, and while I was looking into the process, I decided it was time for a change of pace. In the fall of 2002, I started at the University of Montreal's École Polytechnique, to develop my scientific side by studying engineering. I've always loved engineering: the practical application of math and science to real-world situations appealed to me deeply. From a young age, logic puzzles and math problems had been a favourite pastime of mine, and I relished the opportunity to take on a fresh intellectual challenge.

I also liked that it was entirely unexpected—at least for those who didn't know me well. Ever since the funeral, people had been watching for signs that I was headed into politics, and this unanticipated step was a way of thumbing my nose at them.

While I was a student at Polytechnique, I met Sophie.

IN JUNE OF 2003, I WAS ASKED TO HELP OUT WITH THE Starlight Children's Foundation gala. It was a major production. Tony Bennett supplied the musical entertainment, and Belinda Stronach arrived with Prince Andrew on her arm. I co-hosted the evening with media personality Thea Andrews and a charming Quebec TV and radio host whom I felt I had met before. Her name was Sophie Grégoire, and I caught myself staring at her thinking, Why do I know this woman?

When we finally got to chat, Sophie answered my question. She had gone to school with my brother Michel, and had met me a few times back then. There was a four-year age gap between us, a wide chasm when you're in the teenage years. Still, I remembered her face. And now, of course, the age difference meant nothing.

Sophie had known Michel since the third grade, when they attended Mont-Jésus-Marie school in Montreal, and their paths crossed again at Brébeuf, where she dated one of Miche's best friends. Sophie had regarded Michel as a soft-hearted rebel who loved the outdoors and hated cliques. Brébeuf, as I well knew, could be a snobby place, but Michel had carved out a reputation for himself as a sort of anti-snob.

Five years had passed since Michel died, and the emotional impact of his death had yet to fade for me. (I'm still waiting . . .) But it had healed enough that I could laugh and reminisce with Sophie about Michel's high-school antics without becoming morbid or maudlin.

Sophie and I had a great time that evening, despite difficulties getting some of the inebriated guests to quiet down during Tony Bennett's performance. In fact we bonded over our shared ineffectualness. We spent much of the evening chatting and flirting, and by the time the gala was over I knew she was a very special woman. Then the night was over, and she was gone.

She sent me a quick email a few days later, saying how nice it had been to see me and wishing me the best. I was delighted to hear from her, but was way too chicken to reply. I had already sensed that this was no ordinary encounter and no ordinary woman, and even just meeting for coffee would likely quickly escalate into *the rest of my life*.

I told myself that if it was destiny, then it would happen, and there was no need to rush things. And sure enough, a few months later, at the end of August, I was walking up Boulevard Saint-Laurent when a voice passing in the other direction offered a cursory "Salut, Justin."

Sophie! I wheeled and raced after her, and as she stood there, arms folded, I said the only thing that came to mind: "I'm sorry I didn't answer your email!"

She raised an eyebrow, impressed despite herself that I knew I'd come across as a bit of a heel for ignoring her.

"I'll make it up to you. Let me take you out for dinner," I offered.

"Drop me a line sometime and we'll see" was her casual reply, and she walked off.

It took a few weeks of chatting by email and phone, but

I was persistent. Sophie finally agreed to a dinner date, on the condition that we go somewhere neither of us had ever been. To get outside my comfort zone, I called Sacha for advice, as his tastes have always been even more adventurous than mine. He suggested the Khyber Pass, on Duluth, for Afghan food. Sophie liked the idea, and we were on for the following week. She gave me directions to her apartment, "right in front of the Pierre Elliott Trudeau Rose Garden," she offered helpfully, no doubt with a bit of a wink. I never admitted this to her, but I had to look it up.

I didn't exactly roll up to her front door in a horse-drawn carriage. I had driven a Volkswagen Jetta TDI for some years, a car that had faithfully carried me back and forth between Vancouver and Montreal when I was living in B.C. But the Jetta had been stolen that summer, and I hadn't gotten around to replacing it. Instead, I was driving Michel's old Ford Bronco, which had sentimental value but not much else going for it. After his death it had spent the entire winter buried under snow at the top of the Kokanee Glacier logging road, and no matter how I tried I couldn't get rid of the musty smell it had acquired. Sophie didn't complain, but she did tease me about it.

She and I talked about a hundred different things over dinner, and we always circled back to Michel, my father, and shared memories from the 1980s. Sophie had not only known Michel through their grade-school years; she also had crossed paths with Sacha through mutual friends, often on the ski hill.

It was as if we had lived parallel lives that were now being drawn together at last. It is one thing to be attracted to a woman, to find her witty or poised or smart or beautiful. Sophie was all these things, and both of us delighted in what she now refers to as "the beautiful discomfort" of flirtation. But if the object of your affection doesn't understand what makes you tick, superficial attraction isn't enough. Much of what made me tick was my family, including the father and brother I would never see again. So it's no coincidence that I was strongly drawn to this incredible woman who had known my family in happier times.

The most durable kind of love is woven together from all the things rooted in an intertwined past, including shared values and culture. These are things you don't have to explain, and probably couldn't if you tried. It made meeting Sophie less about getting to know someone new and more like discovering someone you'd both known and dreamed of all your life. And it was why I began to understand over dinner that night that I had returned to Montreal for Sophie, even before I knew her name.

I didn't want to ruin things by chasing her away with overly romantic statements, so I tried to stay generally cool. After dinner, while we were walking along Rue Prince-Arthur's pedestrian promenade in search of ice cream, Sophie said, "Let's go to karaoke! C'mon, it'll be fun!"

The karaoke craze had peaked and most popular bars had sold their machines long ago. But I had seen an Asian place on Rue de la Montagne that still catered to amateur crooners, so we drove there in the Bronco, booked a private booth, and sat together singing the soundtrack from the film *Moulin Rouge*. Sophie's singing voice was excellent; mine not so much. It didn't matter. I was becoming thoroughly enchanted by her, could feel myself relaxing and trusting my feelings in a way that I rarely had allowed myself to before. I felt both vulnerable and safe at the same time, and the confluence of happy emotions had me off balance, so much so that I actually walked into a lamppost outside the bar. (It being a first date, I couldn't convince Sophie that I wasn't actually a klutz; it took years of my never again doing anything of the sort for her to understand the state I had been in that night.)

Back at my apartment, we sat on the couch and talked into the wee hours. The more we revealed about ourselves, the closer we became, and our conversation eventually moved into the realm of sad secrets. Sophie told me about her battle with bulimia as a high-school student and about the loneliness she had experienced as an only child. I in turn told her about my tumultuous childhood.

As our first date drew to a close, I felt a giddy sense that Sophie would be the last woman I ever dated, a feeling that was so strong I actually said, "I'm thirty-one years old, so I've been waiting for you for thirty-one years. Can we just skip the boyfriend/girlfriend part and go straight to engaged, since we're going to spend the rest of our lives together?" The powerful emotions had us both laughing and crying at the same time. The intensity and clarity of that moment left us both at a loss for further words. I drove her home in a loaded, but comfortable, silence.

As I've said a few times when I've told this story, it then took Sophie a few weeks to realize I was serious, and a few more to realize I was right. Yet for me this had been one of those moments of total clarity, in which I had a quiet, unshakable certainty about how things were going to unfold.

My friends and family adored Sophie; I fell in love with her parents as well. The following year Sophie and I bought an apartment together just off Avenue Van Horne, and for the first time in my life, I moved in with my girlfriend. We travelled together; challenged each other physically, spiritually, emotionally, and intellectually; and discovered truths about ourselves through each other.

Sophie is the most colourful, articulate, passionate, and profound spirit I've ever met. Her complex personality is filled with contrasts. An extreme skier able to tackle the toughest pitches and drops, she's also soft, graceful, and maternal. Artistic creativity and a great sense of humour mix well with her strong personal discipline and focus. She's an only child, but has always been curious and drawn to others. Her vulnerability, intelligence, and intuitiveness are exhilarating, and with every passing day I can only love her more.

On October 18, 2004, I took her out to visit my father's gravesite in Saint-Rémi, where I quietly asked for his blessing on what would have been his eighty-fifth birthday, and a few hours later in a beautiful candlelit Old Montreal hotel room filled with rose petals, I asked her to marry me so we could build a life together.

The weekend before, I had visited Sophie's parents at their Sainte-Adèle home in the Laurentians north of the city. As a gesture of old-fashioned respect I actually got down on one knee in the damp fall leaves in front of her father while we were taking a walk through the woods, and asked permission to marry his only daughter. With the characteristic gruff humour Jean always uses to mask his sensitivity, he said, "Yes, of course, of course. Now get up! Your pants are getting wet."

I offered Sophie the ring, in front of the fireplace, in an antique Russian lacquer box Sacha had given me for this occasion. It was a nice gesture, my brother's way of welcoming my choice to ask Sophie to join our family. I'll never forget that moment, in which time was suspended while I waited for her answer. And waited. She was smiling and kind of nodding, with the tears in her eyes matching the ones in mine, and finally, I had to prompt her for an actual answer. Less than a year later, on May 20, 2005, we were married at Église Sainte-Madeleine d'Outremont. We pledged to stand by one another through thick and thin, through bad times and good.

Our marriage isn't perfect, and we have had difficult ups and downs, yet Sophie remains my best friend, my partner, my love. We are honest with each other, even when it hurts. She grounds and inspires me, challenges and supports me. On some days she provides the strength I need to fight; on others she provides the grace I need to stand down. We are blessed to journey together through this life. Given the changing fortunes of time, our love is what reminds us of what really matters.

After a couple of years at Polytechnique, I recognized my studies for the intellectual indulgence they were. I never intended to become a professional engineer, and I realized that my other involvements were not just taking up more of my time but were more consistent with my primary skills and interests.

I was by this time chairing the Katimavik board, a position that included successfully encouraging the Liberal government of Jean Chrétien to increase and stabilize funding to $20 million a year, and cross-country speeches to high schools on the value of community service and volunteerism.

I was also on the board of the Canadian Avalanche Foundation, where I promoted avalanche safety through events at ski resorts across the West, pressured provincial governments in B.C. and Alberta to help fund the Canadian Avalanche Centre and its public advisories, and helped with private fundraising for the organization. Indeed, it afforded me my first opportunities to understand how philanthropically minded westerners are, as our annual fundraisers at the Calgary Zoo were extremely well attended by leading figures of the Alberta oil patch wanting to contribute to a worthwhile cause.

For a year I had a weekly segment on French radio with CKAC, covering current events (and being their official 2004 Olympics correspondent from Athens), which gave me the chance to get to know the Quebec media and cultural scene from the inside. It also taught me how powerful radio can be as a means of connecting with people. You can't be phony on radio: your voice and tone will give you away. And people don't care what your name is after the first ten seconds. All that matters is what you have to say, how you say it, and that you're speaking *to* people, not *at* them. Which is partly why to this day my favourite interviews to give are on the radio, in studio, engaging live with my interviewer.

As well, the Canadian Parks and Wilderness Society brought me in to head up their Nahanni Forever campaign, to protect and expand the Nahanni National Park Reserve in the Northwest Territories. After a canoe trip down the Nahanni River with various environmentalists and journalists, I embarked upon a national speaking tour to promote the CPAWS campaign.

In fact, I was getting called upon more and more often to speak at conferences and various events on youth or the environment, and although my teaching background and work with Katimavik gave me expertise on the former, I recognized that I needed a deeper understanding of environmental issues. And so in the fall of 2005 I decided yet again to continue my education, this time at the graduate level in environmental geography at McGill.

It was also that fall that I approached a speaking agency

to help me manage the requests I was getting. I had so far resisted the idea of charging speaking fees, but I needed help with arrangements and logistics. I also began to understand the market forces around public speaking, especially in regard to fundraising. The right speaker helps to fill a room and sell out tickets for a charity event, sponsors are more than happy to help in exchange for visibility at a successful community event, and a good speaker helps set the tone and frame the success of a professional conference. For many conferences, hiring a speaker is part of the budget, just like renting a hall, catering, or booking musical entertainment.

Of course, I continued to speak pro bono for a number of causes I was engaged in, from Katimavik to winter sports safety to protecting the Nahanni. And whenever possible as I travelled across the country, I got in touch with local schools and offered to do a free event with them during the day, since I was in town in any case for a paid event.

The more I spoke with young people all over the country, the more I began to gravitate toward a life of advocacy. It was becoming increasingly clear to me that the issues youth cared about—education, the environment, their generation's economic prospects—needed a stronger voice in the public sphere. I also began to feel that a generational change was approaching, one that might open up new possibilities. It was against this backdrop that I made my first steps into politics.

FOLLOWING THE LIBERALS' JANUARY 2006 ELECTION loss, Paul Martin stepped down as leader, and by that spring an eleven-person leadership race was well under way. I chose to stay away from it, but I did wonder whether, given my growing experience as a public speaker on youth and the environment with a message about citizen engagement, I might have something to offer a Liberal Party in renewal. I talked it over with Sophie, because it would be a big step with possible far-reaching consequences, but we both agreed I had something to contribute and therefore I should at least offer my help.

I didn't know where to begin, but I had heard that Tom Axworthy, whom I had come to known slightly over the years because he had been one of my father's advisors, was heading up the party's Renewal Commission. I called him up and offered to help out with youth issues. The commission was hoping that while much of the party was wrapped up in leadership strategies, a number of people would hunker down and build a tool kit of fresh ideas, policies, and principles that the next leader would be able to draw on to rebuild and renew the party.

That summer my colleagues and I travelled across the country, listening to young people's views on politics and on the Liberal Party in particular. Our goal was to produce a report that would recommend how the party could prompt young Canadians to vote Liberal. But after hearing from hundreds of young people, I concluded that our most pressing challenge wasn't persuading them to vote

Liberal; it was getting them to vote at all, for anyone. In our report, my colleagues and I proposed that the party's primary goal should be to overcome the apathetic attitude of young people and persuade them to participate in elections. Whether or not they would be motivated to choose the Liberal Party once they were in the voting booth was up to the party and the local candidates.

There are plenty of passionate young activists in Canada. However, most of them focused their efforts on performing work with non-governmental organizations, rather than with political parties. "Youth prefer to take individual steps toward making a difference in society," we reported, "but have less faith in the ability of collective efforts to make a difference such as participating in democratic or government initiatives." We noted that young people were committed to environmental actions such as recycling their own garbage, but not nearly as committed to involving themselves in elections, even to the point of taking such basic steps as casting a vote. When you worked with community organizations, NGOs, or even big single-issue causes, it was easier to feel you were contributing in a small but meaningful way to changing the world. When you voted in an election or worked on an election campaign, you were participating in a system in a way that might, abstractly, one day lead to change, but it was far from certain—particularly given the cynicism about politics that was dominant at that time. "Continuing decline in voter turnout will only worsen unless young people are engaged," we concluded.

Among our recommendations, we suggested that politicians engage youth by focusing on issues of importance to younger Canadians, including education, the environment, foreign policy, and the protection of individual rights. We also proposed promoting "a culture of responsible citizenship" by expanding our national commitment to youth volunteerism, and urged Elections Canada to work with high-school boards in conducting mock elections on the same day as real federal elections.

I thought then, as I do now, that citizen engagement is both an end in itself and a necessary means to solve the problems we face as a country. We have some big issues to deal with, and I often worry that unless we reinvigorate our democracy, we will never find legitimate answers to them. Modern democracy shouldn't just be about citizens endorsing a vision and a set of solutions with their votes, but about actively contributing to building that vision and those solutions in the first place. This is the heart of the matter when it comes to democratic reform. Too often reform gets depicted as an "inside the bubble" issue that only politicians and Ottawa people care about. That misses the point. The people who feel the consequences of our democracy's failures most acutely are physically and metaphorically a long way from Ottawa.

I was just beginning to understand the importance of this issue as we published our report in the autumn of 2006. I met with some contenders in the leadership race to hear their opinion of our recommendations and to gauge how seriously they were taking the problems that the party was facing with young voters. I also wanted to get a sense of who really understood the need for real renewal and the opportunity that an election loss gave for modernizing the style and approach of the party. Party members and journalists had been asking me for my view on the leadership contest for some time, and I wanted to better know the field of candidates before expressing any opinion. My own belief was that the party needed to make a break with the bad habits of the recent past and that it was crucial to move away from the sense of entitlement that came with thinking of the Liberals as the "natural governing party" of Canada.

Ultimately, I chose to support Gerard Kennedy, the Ontario education minister. I was impressed with Gerard Kennedy's long record of public service outside government, something many career politicians lack. For example, having run Toronto's Daily Bread Food Bank for almost a decade, he understood poverty, income inequality, and unemployment, issues I was beginning to focus on in my own political thinking. I became enthusiastic about his beliefs and his achievements, his focus on grassroots renewal, and his obvious work ethic. I felt even then that the Liberal Party was in a deeper hole than many within the party realized, and it would take a leader from a new generation, someone from outside the federal party, to reinvigorate it.

For those looking to breathe fresh life into the Liberal Party, the December 2006 leadership convention in Montreal was an inspiring event. Far from either a coronation or a duel between feuding party elders, it was a raucous, unpredictable

nail-biter featuring four candidates—Michael Ignatieff, Bob
Rae, Stéphane Dion, and Gerard Kennedy—with legitimate
shots at victory. The others—Ken Dryden, Scott Brison, Joe
Volpe, and Martha Hall Findlay—all brought enough support to the convention to have a bearing on the outcome.

Looking back at my transition into political life, that
weekend in Montreal really mattered. Until that point,
although I had spent some time working at the margins of
the Liberal Party, I was not yet convinced that I was interested in a career in politics. I loved the world of ideas, values,
and policy-making that lay at its core, but my mother had
warned me, with her words as much as by her example, of
the incredible personal costs to a politician's life. And there
was of course another consideration: entering politics at the
federal level would suggest I was following in my father's
footsteps, perhaps even harbouring the notion that as the
son of Pierre Elliott Trudeau, I somehow deserved a role
based on that qualification alone.

The association with my father was never a reason for
me to get into politics. It was, rather, a reason for me to *avoid*
entering the political arena. The battle to convince myself
and others that I was my own person had challenged me all
through high school and university. Why should I negate those
efforts by making the one career choice that would guarantee
I would be measured according to my father's achievements?
It made sense for me to stay out of that arena for at least
another decade and reduce the inevitable comparisons. That
was my state of mind as the convention approached.

Things changed soon after the gavel was struck in the Palais des congrès. As I mixed with hundreds of other Liberals all intent on the future of the party and the country, I began reassessing my fear about comparisons with my father. Perhaps, I thought, I had underestimated the very real differences between myself and my father when it came to politics.

From the beginning of his political career, my father assumed an intellectual approach to all his political activities, including campaigning. He felt somewhat unsuited for the baby-kissing aspect of electioneering and avoided so-called retail politics whenever possible. Busying myself on the convention floor revealed to me that where political campaigning was involved, I wasn't at all my father's son—I was Jimmy Sinclair's grandson. Grampa Jimmy had perhaps been the ultimate retail politician, a man who loved mixing with people, shaking hands, listening, and, yes, when the opportunity arose, kissing babies. The contrast between the two men is dramatic, and the more it became clear to me, the more it eased my concern about being compared with my father.

I was surprised and enthused by the response I got from party members on the convention floor. Kennedy organizers had to create an advance team for me, to ensure I could move smoothly through the crowd. I genuinely enjoyed working the room for Gerard, discussing issues with delegates and bonding with fellow Liberals. I made a brief introductory speech on his behalf, helped him with his own address to the delegates, and then settled back to watch the outcome of the race.

The vote count on the first ballot saw Mr. Ignatieff cruise to first place, with 1,412 votes. Just 123 votes separated the next three: Mr. Rae with 977, Mr. Dion with 856, and Gerard two behind with 854. Mr. Brison, Mr. Volpe, and Ms. Hall Findlay dropped out of the race, leaving about five hundred votes to spread around in the second ballot and ensuring that the top four contenders were still very much in the race.

The second ballot proved disastrous for Gerard. He remained stuck in fourth place, having picked up just 30 new votes. When Mr. Dryden, in fifth place, was forced to withdraw, he announced he was supporting Bob Rae and freed his own delegates to vote their choice. Gerard voluntarily withdrew and moved his support to Stéphane Dion. I had already decided that should Gerard not be the winner, I wanted it to be Mr. Dion, so I went to him as well. He was a Quebecer who was a strong and thoughtful federalist. Moreover, he had built his campaign around environmental policy, which aligned with so much of what I had heard from young people as chair of the youth task force. Most of all, he was a serious guy. He thought things through deeply and addressed complex issues earnestly. I still find that enormously appealing about Stéphane Dion.

On the third ballot, Mr. Dion almost doubled his votes, leaping ahead of both Mr. Rae and Mr. Ignatieff. When Mr. Rae was forced to withdraw, he released his delegates to vote as they chose, and the result was dramatically revealing. The chasm of support between Mr. Rae and Mr. Ignatieff was so wide that the vast majority of Rae delegates moved to Mr. Dion, who took the leadership on the next ballot.

The day after the convention ended, I phoned Stéphane and congratulated him on his victory. I let him know how happy I was to have contributed to the start of the party's rebuilding. "But now I'm going to step away for a bit, try to get back into private life," I told him. And Stéphane replied, "Don't go too far, because I'm going to want your help in getting rid of this Harper government."

For him it may have just been a polite remark, but after I hung up the phone I looked across at Sophie and told her what he'd said. We realized we had a big decision to make.

THE EXPERIENCE OF THE CONVENTION HAD TAUGHT ME something: I had political skills independent of my last name. I'll not pretend the name didn't make a difference, but it wasn't all I had. Not by a long shot.

The next few weeks I spent in deep discussion with Sophie about the challenges, sacrifices, and opportunities that were part of political life. I consulted with friends and family to get their advice, and I thought long and hard about the effect it would have on all of our lives. But I sensed that the timing was right for me.

I had finished the coursework for my master's in environmental geography, and only writing my thesis remained. If my foray into politics didn't work out, I would be able to pick up where I left off.

Jean Lapierre, Paul Martin's former Quebec lieutenant and MP for Outremont, had announced after the convention that he would not run again. Encompassing the north and east flank of the mountain that marks the centre of Montreal, Outremont held the Trudeau family roots and my seven years at Brébeuf, and it's where Sophie and I bought first a condo and subsequently the house we were then living in. The riding was a natural fit.

On top of that, I knew that it was no longer the easy riding that it had always been. So I would have to work hard to win it, after winning over all those Liberals who didn't bother to hide their opinion that I hadn't yet paid my dues to the party. And I knew that it would be through that hard work that I would demonstrate to all that I was more than just a last name.

So a few days before Christmas, I called Stéphane again and told him I was interested in running for him, and that I thought Outremont would be a good fit. He thanked me and told me he'd get back to me.

A week or so into the new year, Jean Lapierre announced that he would be stepping down immediately, and therefore there would be a by-election in Outremont. Perfect, I thought, somewhat naively. It'll be a tough fight with lots of attention, and we'll really show that the Liberal Party is serious about generational change.

Within days, however, the level of internal Liberal intrigue around the riding had escalated from the standard "merely unpleasant" to "outright toxic," and it was made known that the Outremont riding association vehemently opposed even the rumour of my running there. The leader's office wasn't so keen, either. So for me it became a clear no-go.

I wasn't as put out as I might have been. Indeed, following my brother's advice, I had already begun to look around for other suitable Montreal ridings to run in, and two stood out clearly: Papineau, farther north and east, around Parc Jarry, and Jeanne-Le Ber, south of downtown in Verdun. Both were diverse urban ridings with significant economic challenges. Most important, both had been won by the Bloc Québécois the year before, so a win in either would not just be holding a Liberal seat but would be taking a riding back from the sovereigntists. What better way to prove my worth?

Of the two, Papineau just felt right. I already knew well the great range of ethnic restaurants in Parc-Extension, had been to friends' weddings in the Orthodox churches, had enjoyed memorable sunny days in Parc Jarry and, like many Montrealers, had shopped for curtains along Saint-Hubert. Walk the length of the riding along Rue Jean-Talon and in addition to French you'll hear English, Greek, Punjabi, Bengali, Tamil, Urdu, Spanish, Portuguese, Arabic, Creole, Vietnamese, and Italian. I could keep up my world travels simply by crisscrossing the riding!

On the cusp of Montreal's east end and near the geographic centre of the island, Papineau borders Outremont to the south and my father's old riding of Mount Royal to the west. At just nine square kilometres, Papineau covers the smallest area of any federal riding in Canada. According to the 2006 census, it had the lowest average family income of any constituency in Canada, and although it has as wide a range of language and ethnic representation as you could expect to find anywhere in the country, it is also solidly and indubitably francophone first. Although Liberal for many decades, Papineau had fallen to the Bloc Québécois in 2006, when a Haitian-born star candidate named Vivian Barbot defeated Liberal foreign affairs minister Pierre Pettigrew in a close race.

It was the ideal setting to prove my electability. Papineau was just the sort of vibrant, multicultural riding the Liberals needed to win if they were going to become competitive in major urban centres in the next election. It had the francophones that Liberals would need to win over to regain Quebec. And its compact size suited an energetic, shoe-leather candidate willing to knock on doors from one end of the riding to the other.

But then Dion's people indicated to me that Papineau was not right for me either, as they had earmarked it for an "ethnic" candidate. Clearly trying to dissuade me, they told me that it was going to be an open nomination, which suggested to me that they thought it unlikely I could win a contested nomination race.

It soon became clear that Dion's team was leaning toward the locally beloved Mary Deros, who had represented Parc-Extension on city council since the late 1990s. I suspected they saw me as someone who wouldn't be willing to put in the elbow grease it would take to win such a challenging riding.

None of the obstacles in Papineau discouraged me. On the contrary, I considered it an excellent opportunity to prove myself in a difficult situation. I didn't want a cakewalk. I wanted a serious test of my political abilities.

But as much as winning the nomination was going to be about my individual political abilities, I knew that I needed to show from the start that I was a team player. Loyalty and respect for the leader was something I believed in deeply; indeed, I believed it was key to the party's regaining the respect of Canadians. So when, one morning in late February 2007, I heard that a rumour was beginning to circulate about my interest in Papineau, I immediately asked Dion's team for guidance on how to handle it. I knew that it was just a matter of time before a reporter asked me directly, and it would be big news when I confirmed, so I felt a need to coordinate with Liberal Party communications.

The answer I received was straightforward, and somewhat dismissive: answer journalists' questions however I chose. Furthermore, they gave me the distinct sense that I was a bit full of myself for contacting them about such a piddling matter. I sighed. I obviously had much to learn about politics.

Within hours, a Radio-Canada reporter called me at home to find out if, indeed, I was planning to run in Papineau. When I said that I was, she asked if they could interview me on camera. I said fine, but it would have to be at the airport, as I was on my way out west for an avalanche safety event.

When I arrived at the terminal, a number of cameras had gathered from various outlets, and I gave a quick press conference. I then boarded my flight for Vancouver, feeling that all had gone quite well.

When the press followed up with Stéphane Dion, he confirmed my decision, told reporters he admired my courage, and proclaimed that I was proving I wasn't taking "the easy way" toward election as an MP.

But by that evening, his people were livid. Mr. Dion had been in Montreal the same day, delivering a major speech on terrorism, which the media barely touched on. The news was all about my decision to run. Over the next days, many senior Liberals accused me publicly of deliberately upstaging Mr. Dion just as he was trying to get his footing as the party's new leader.

It was for me a tough but illuminating introduction to the internal workings of the Liberal Party of Canada, where infighting, personal agendas, and lack of coherence were the norm.

But far from turning me off from the hard work that lay ahead, it had simply whetted my appetite. Bring it on, I thought. This is gonna be fun.

CHAPTER SIX

Papineau: Politics from the Ground Up

═══

MY CAREER AS A POLITICIAN BEGAN IN A parking lot. A grocery store parking lot, to be precise, directly across the street from a shawarma restaurant and a barbershop. The cameras and reporters who had come rushing to the airport to breathlessly relay that I planned to seek the Liberal nomination in Papineau were nowhere to be seen. Now it was just me with a clipboard, approaching strangers to ask if they would pay ten dollars to purchase a Liberal Party membership. Welcome, I thought, to the glamorous world of Canadian politics.

This wasn't the actual election campaign. It was the opening days of the nomination battle to choose the candidate who would carry the Liberal banner in Papineau once the election was called. I was in the fight armed with limited money, barely any retail-politics experience, a couple of friends as volunteers, and a staff of one, who happened to be my wife. Sophie was full of enthusiasm and support for me, helping plan the approach and joining me on the ground from time to time, both of us learning the ropes together.

Most Canadians are unaware of the clashes that can occur during the party nomination process at the constituency level. It's behind-the-scenes stuff that remains hidden compared with all the hoopla generated when an election is in full swing. In some cases these contests are bypassed when incumbents and high-profile challengers win their nominations unopposed. But for would-be candidates involved in a battle to secure nomination votes from party members, it can be a gruelling contest. It begins with each candidate recruiting as many members as possible in advance of the nomination meeting, then inducing these members to show up at a community hall, school, or arena on nomination day to cast their vote. It may sound like drudgery. But *I loved it*.

After experiencing the drama and manoeuvring of the 2006 Liberal leadership convention, this brand of up-close-and-personal politicking quickened my pulse. I'm a social being by nature. I'm also someone who enjoys physical activity, which campaigning in Papineau required in large degree. Walking the streets of the riding from dawn till dusk to sign up members was enormously appealing to me, and I could hardly wait to get started each day. I understand the importance of working the phones, but all things considered, when it comes to campaigning I prefer to wear out a pair of shoes, meeting people and getting things done at street level.

The work was rewarding for another reason. The grassroots of the Liberal Party had shrivelled through a combination of hubris, overconfidence, and neglect. In many regions of Canada, some Liberal candidates didn't even bother to walk the neighbourhoods and knock on doors; they considered the Liberal Party more of a brand than an expression of political vision. This attitude was at the core of our reduced support. We needed to move beyond that kind of thinking. We needed to remind voters of the values and philosophy behind Liberal Red. More important, we needed Canadians to remind *us* of their hopes and expectations for their community and country. That may sound like an obvious thing, but it's amazing how often and how easily people in politics forget it.

It is often said that politics is a "contact sport," by which people mean it is tough business. That's true. It is not for the faint of heart or thin of skin. But I think of it in another way: politics is a tactile business. You need to spend time, real time, with the people you seek to represent: in coffee shops, around kitchen tables, at backyard barbecues. You have to listen, and absorb the views and values of your community. You have to work at it. I wanted to instill this as a core ethic over the course of my campaign in Papineau, both for its own sake and because the Liberal Party needed this kind of rejuvenation. Badly. As a rookie candidate, I couldn't do much about this nationally, but I could locally, and I set out to get it done with a direct, vigorous, and personal way of delivering the message. I knew this was the right way to do things.

I knew this approach would be particularly important in my home province of Quebec. The sponsorship scandal and consequent Gomery Commission had painted a gruesome picture of the Liberal Party for my fellow Quebecers, and it stung. Perhaps the main reason I had supported Gerard Kennedy in the 2006 leadership campaign was that, as an outsider to federal politics, he seemed to understand just how disconnected and imperilled the party had become. The party's basic integrity had been called into question, and I was convinced that the only way to fix that was the old-fashioned way: look people straight in the eye, listen to them, and tell them the truth.

Of course, enthusiasm and good intentions can take you only so far. In Papineau, they had brought me as far as this grocery store parking lot. The experience stood in stark contrast to being on the floor at the leadership convention elbow to elbow with several hundred Liberals. Many of them had been supporting candidates other than my own, but we all shared the same party identity and ultimately the same goal. On the streets of Papineau, I could never assume anyone's politics. In fact, I could never assume they would respond to my greeting and invitation with anything more encouraging than a quick smile and a shake of the head.

Whenever I could persuade someone to stop and discuss things with me, I would talk about my concerns about Stephen Harper's approach to leadership and the way his party was running the country. I would explain the ideas on education, youth involvement, volunteerism, and the environment that I had developed during my years as a teacher and as chair of Katimavik.

Most important, I did a lot of listening. The only real way to expand my understanding of the issues voters were facing was by asking them what concerned them and listening carefully to their answers. I heard parents tell me how hard it was for their children to land jobs, I heard immigrants describe the difficulties they were having securing visas for visiting relatives, and I heard about the economic challenges that many of the residents faced day to day. Their debts were growing but their incomes weren't. Many of the shoppers emerging from that grocery store could barely afford the food to feed their families.

I also learned about some of the concerns that some residents had about the changing nature of the riding. Greek and Italian immigrant communities in Papineau were well established and accounted for much of the area's vitality. The influx of new arrivals from other nations was leading to a blossoming of ethnic restaurants, festivals, and community centres, which elevated Papineau's overall liveliness. Many long-time residents, however, told me they worried about friends and neighbours being crowded out by the newcomers.

The range of cultural viewpoints constantly amazed me. Some residents reminded me that Villeray, the increasingly Latino neighbourhood at the heart of Papineau, traced its roots back to French-Canadian farmers and quarry labourers who first worked the land in the days of horse and buggy. Greeks in Papineau's Parc-Extension neighbourhood were watching their children and grandchildren move to distant off-Island suburbs, while their houses were being bought by immigrants from South Asia, and in the traditionally Italian Saint-Michel neighbourhood, newcomers from Haiti and North Africa were moving in. Added to this mix were the job challenges, not just for newcomers but for young people as well. Communities all across Canada have been similarly transformed in recent decades, but Papineau's multicultural mosaic was far more complicated than most. I loved the vivacity of the riding, but grew concerned about the tensions developing.

I was battling two other candidates for the Liberal nomination in Papineau: the party leadership's preferred choice, Mary Deros, and Italian-language newspaper publisher Basilio Giordano, who also had the backing of influential Liberals. Both had well-organized political machines that successfully recruited large groups of members by working with local community leaders. Lacking those kinds of connections, I had to work at bringing in members by ones and twos while my competitors corralled them by the dozens.

To make matters more difficult, the president of the Quebec wing of the Liberal Party announced to French media that I had no business becoming a candidate, as I had nothing to offer. And it was not just the Liberal establishment that wrote me off before I even started: political pundits and columnists announced to the world that I had shown already that I was out of my depth by picking a riding in which I had no chance of winning even the nomination, let alone an election against a star like the Bloc Québécois's Barbot. My inevitable failure, they explained, proved that I was young, foolish, and obviously not half the politician my father had been.

Indeed, the math didn't seem to be pointing in my favour, and with two months to go before the nomination meeting, it became clear that I needed help. It arrived in mid-March when a dear friend, Reine Hébert, agreed to join our two-person team. Reine was a veteran Quebec-based political campaigner who had worked with the federal Liberal Party since my father's era. I also recruited Franco Iacono as my campaign director, and together they helped maximize my exposure to Papineau residents. With their help, I managed to track all the one-time Liberals in the riding who had let their memberships lapse. We visited them at their homes and encouraged them to rejoin the party, and though a few weren't interested, many liked what I had to say and came aboard. By April 29 we had sold almost twelve hundred Liberal Party membership cards, about the same number as the other two candidates. It was still anybody's race. The winner on nomination night, it became clear, would be the candidate whose speech won over enough of the members who had been recruited by the other two nominees.

As the meeting date approached, I grew increasingly optimistic. At the end of April, I started to win over some critics, largely because they recognized the sheer dogged-ness with which I worked the riding. I also believe I had better insight than my competitors into the changing nature of politics and media. When a local blogger asked each can-didate a series of questions about poverty, identity politics, immigration, and other issues, I responded with lengthy, personal replies that drew on my experiences in the riding. The other candidates chose not to respond to him at all, presumably assuming that few voters bother to read polit-ical blogs. But even in 2007, I knew that the Internet was becoming a critical tool for expanding a political party's out-reach, especially to younger supporters.

To my delight the blogger, who had been telling his readers I was destined for failure, gave me a respectful nod on his website for taking the time to answer his questions. This was a small thing—I doubt that it swayed more than a handful of votes—but it reinforced my belief that today's activists and supporters expect and deserve direct engage-ment through the digital media.

The nomination meeting was held at Collège André-Grasset, just across the riding boundary. Despite the unpredictability of the outcome, I felt calm when I entered the auditorium, comforted by the presence of the people I loved most in my life. My mother was there with Sophie, and my brother arrived carrying his four-month-old son, Pierre. Sacha was busy with his young family and his career as a documentary filmmaker, so I greatly valued his support.

From the beginning I had been identified as the underdog in the race, and this helped relieve some pressure on me. I had entered a tightly contested race against two experienced candidates. If I made it to the second round of balloting, many observers predicted that the Deros and Giordano supporters would combine forces to defeat me, which seemed likely. This was, after all, the hard-knocks school of politics. I wanted to win, but under the circumstances there would be no shame in losing.

Looking out over the crowd that evening, I was struck again by how different I was from my father when it came to this kind of thing. He hadn't spent a lot of time hanging around grocery store parking lots meeting voters or fighting nomination battles in school auditoriums. Of course, politics was different in his day. Back then, high profile candidates were drawn from elites such as bankers and lawyers. They earned the confidence of voters through their stature in the community, with a perspective extending far beyond their own riding to encompass all of Canada. My father suited that description perfectly. He thought of himself primarily as someone who represented Canadians and their values in a broad national sense, as opposed to representing one riding.

He was a good MP, and the issues he fought for were consistent with the interests of the voters in his riding of Mount Royal. But he didn't aspire to have the *personal* connection with voters that I was determined to forge on the streets of Papineau.

With that in mind, I had planned to avoid any mention of my father in my speech that night. I believed it would only hurt my cause if the members suspected I might be trying to ride on his coattails.

Reine advised otherwise. She reminded me that parts of the modern Papineau riding were once within the borders of neighbouring Mount Royal, which meant at least some voters in the room had been represented in Parliament by my father. Failing to at least acknowledge this fact might strike these people as being disrespectful of him. And besides, she told me, if you do this right, this one time, it'll be a long while before you need to bring him up again. Her words guided me as I prepared my speech.

I began with a short nod to history. "In the fall of 1965," I said, "the residents of Parc-Ex helped send Pierre Elliott Trudeau, who listed his occupation as 'teacher,' to the House of Commons for the first time. Times change, and riding borders change, but what you were part of forty years ago changed Canada forever." I reminded them that it had been twenty-five years since my father had given Canada the Charter of Rights and Freedoms, one of the most valued tools the world has ever seen for ensuring the protection and full exercise of human rights. "Now we are all children

43. Here I am showing caucus colleagues that yoga can really be done anytime and anywhere.

44. Papineau's diversity never ceases to amaze me, and there is no better opportunity to get out and see it than during campaign time, as shown here at a local Portuguese club during the 2011 federal election. With endless political discussions and many cultural celebrations, Papineau is a true testament to Canada's own diversity.

45. The Papineau campaigns defined how I saw the direction the party needed to go after the 2011 election, with a strong ground game and continuous community engagement within every riding.

46. Election day in the riding means not only getting out the vote but also making it out to all the polls and thanking the staffers at the voting stations for their work.

47. I thought it would be hilarious to stick my face in one of my defaced signs. You can't get too personally upset by vandalism or attacks; it's easier to have fun with it instead.

48. Ali Nestor Charles, who spars with me here, is a boxer I met through his work with street youth and community outreach in my riding.

49. Meetings and briefings can happen anywhere, as you can see with this greenroom-bathroom hybrid before my leadership campaign launch—with Gerry Butts, Katie Telford, and others.

50. Sophie weighs in very strongly on many of my speeches. From the tone to the approach to the ideas, she is a vital partner for me, ensuring I hit the right notes in both substance and language. She is a brilliant and invaluable sounding board for me in all that I do.

51. The debates were my first opportunity to directly engage in the give and take of ideas at the highest level. I enjoyed challenging others almost as much as I enjoyed being challenged in return.

52. A leadership campaign stop at a library in Brantford, Ontario.

53. Goofing around and sword fighting with the kids minutes before my speech at the Liberal leadership showcase in Toronto, while Gerry monitors his Twitter stream.

54. This photo is all thanks to Alex Lanthier, who was dancing ahead of us before the leadership showcase speech. He's always been a great dancer and knew when to help us relax before a big event.

55. The last few quiet moments before the showcase as Sophie calms and grounds me.

56. After various drafts go back and forth, Gerry and I input my last changes to speeches, as we are doing here while waiting for the results from the leadership race.

57. The kids had been looking forward to the leadership results all night, but between their not being able to read the results and my not reacting, it took them a little while to catch on. A family photo, really, with my family, Sophie's parents, my mom, and my sister, Ally, all in the shot.

58. My first media scrum as the new leader of the Liberal Party.

59. Also on my first day as leader, I reached out to the Liberal provincial premiers from my new office.

60. Connecting with kids is always easier if you read them a story to make them forget the mob of people who barged in on their class. This was a great group of kids in Brandon, Manitoba.

61. Work on the Hill is more than just Question Period and caucus meetings. Meeting visitors and constituents is always a good reminder of why we are there in the first place.

62. It is important that constituents be able to connect with their MPs and see them at work. Here are Papineau residents visiting me in Ottawa for the day to see the other aspect of my job, away from the riding.

of that Charter," I said. "Of that we are immensely proud. So you can understand how fiercely proud I am to be able to say that your Prime Minister Trudeau was also my dad."

I went on to speak from the heart about Papineau itself, naming and congratulating the community leaders who were the heart and soul of its neighbourhoods and cultural communities. Then it was time to cite the Conservative government policies that I would oppose as a Liberal MP. "The Conservatives want to divide us on social justice. They want to divide us on the environment, on Kyoto, endangering the future of our children. They want to divide us on Canada's role in the world, with positions copied from the American right."

For two months, I had been up and down every street in the riding, visited every mall and shopping area, shaken hands with thousands of people, and listened to innumerable stories. I felt I had gotten to know the people of this riding and their concerns. No matter how the votes went that evening, no one would be able to say that I hadn't worked to win the nomination.

But as I delivered my speech to the crowd, something began to happen. Wherever I looked, I saw familiar faces looking back at me, people I had stopped to chat with on the sidewalks of Saint-Laurent, Saint-Hubert, and Christophe-Colomb, or whom I had met on their front porches along Rue Everett and Avenue de Chateaubriand. I had connected with these people. They smiled at me. More important, they supported me once balloting began.

I won on the first ballot that evening, taking 690 of 1,266 ballots cast, versus 350 for Ms. Deros and 220 for Mr. Giordano. When the numbers were announced I gave Reine a huge hug, then looked over to see Sacha blinking back tears of joy at my victory.

In the midst of our celebration, Stéphane Dion telephoned to offer his congratulations. He and his team might have favoured Mary, but now that the decision had been made he didn't hesitate to offer his sincere congratulations, and I told him that I greatly appreciated his gesture. Our relationship had been neither warm nor antagonistic during the nomination race. Stéphane had treated me with nothing but respect, and during the politically turbulent year and a half that would follow my victory, I was careful to show him the same respect in return.

WHILE CELEBRATING MY VICTORY THAT NIGHT, I KNEW my fight had just begun. This had been, after all, a skirmish among friends. The real battle would take place when I went up against the incumbent MP, Vivian Barbot. Mme Barbot had forged an impressive career as an educator, feminist, and leader of the local Haitian community before winning the Papineau riding by defeating Pierre Pettigrew, a prominent Liberal cabinet minister. She was a more than worthy political opponent, and I knew there was no time to waste if I wanted to defeat her in the next election.

My first step was to create an organization, a network of

people with the experience and insight to run an effective campaign. Reine and Franco had been invaluable to me, but they were moving on to other commitments. Family and friends had provided plenty of help in the final weeks of the nomination run, but I couldn't expect them to be available through the marathon of a federal election.

The first position I needed to fill was that of a full-time campaign manager, and I discovered that the ideal person had been in the room with me the night I won the nomination. Louis-Alexandre Lanthier had been brought in by Reine to act as my scrutineer. I had briefly met him in Ottawa during my Katimavik days. Still in his thirties, Alex was a veteran of many political campaigns, following in the footsteps of his mother, Jacline Lanthier, who had worked for Jean Chrétien. I liked Alex's background and his commitment to making the upcoming election a success for us. I especially liked the way he summed up the strategy for the upcoming campaign in a single sentence. "Justin," he said, "until the next election, you're going to act like you are *already* the MP for Papineau."

It was an extension of the strategy we had pursued successfully during the nomination battle. At every restaurant opening, every religious festival, outdoor carnival, fundraising event, parade, fashion show, arts exhibition, bazaar, town hall meeting, or any other public event taking place in Papineau, I would be there shaking hands and talking to people. I wouldn't be asking for their vote—the election date hadn't been set. I would be there to let people know who I was, to put a face and a voice to the name they would see first on campaign material and later on the ballot. When the time came to decide how they would cast their vote, they would remember the guy who shook their hand and who asked how they were doing and who wanted to hear their concerns. It was perhaps the ultimate form of retail politics, and I could hardly wait to get started.

Alex's grandfather had run a laundromat in Papineau for many years. A lot of people in the neighbourhood were unable to afford their own laundry appliances, so the business became something of a local institution where people would gather to chat and knit and read and, almost incidentally, get the family's laundry done. I couldn't think of better grounding for someone assigned to familiarize the voters with me and me with them.

From spring 2007 until the general election of October 2008, my days were consumed with showing up at events and talking to people. I can't say that I met every single voter in Papineau during that year and a half, but if I didn't it wasn't for lack of trying.

Wherever I appeared and whomever I spoke to, I made strategic decisions about the way I communicated my message. Identity politics might have been one way of establishing rapport with voters, the kind of divide-and-conquer strategy favoured by the other parties. But I had no intention of going down that road. I tried to build common ground around common values that I believed were widely shared in the riding. No matter where the people came from, what language they spoke, or how they prayed, I believed we all shared certain values, and I wanted to emphasize this connection between us.

The largest ideological sticking point I faced, particularly in the Villeray neighbourhood, dealt with the question of Quebec's status within Canada. I'd regularly knock on a door, watch it open, only to see a face visibly blanch upon recognizing me. "I'll never vote for you," they'd start. "I don't agree with you."

"Ah . . . ," I'd reply with a friendly smile. "You don't agree with me on the environment? Or on social programs, education, health care? On an economy that provides opportunity for everyone? On disagreeing with Mr. Harper?"

"No, no, we agree on all that. We just disagree on Quebec!"

"Well, okay, except that we don't *actually* disagree on the need to protect the French language and culture here in Quebec. Where I suspect we may disagree is in the best way to do that: I believe in strengthening and sharing our language and culture. I don't believe in turning inwards and building walls."

However the conversation went, I'd let people know that I was glad to have met them and that I hoped to be a strong voice for them in Ottawa on important issues, whether they voted for me or not.

I was always careful not to disparage the Bloc MP, Vivian Barbot. I had a good deal of respect for what she had accomplished in her life. Personal attacks are not the way to win a voter's allegiance.

I became convinced that the only way to fight the sovereigntist perspective over the long term is to make the positive case for Canada. Like other divisive forces within the
country, the sovereigntist attitude is fundamentally a choice
to emphasize the few things that divide us rather than the
many things we hold in common. When you scratch the sur-
face on the streets of communities across Quebec, you soon
find those common values and aspirations. As I would say a
few years later when launching my leadership campaign, I
share those values, but strongly believe that Canada, not just
Quebec, is the best place to translate them into reality.

I was able to bond with just about everyone I met in
Papineau, sovereigntist or otherwise, by discussing my oppos-
ition to Stephen Harper's agenda. In the short time since they
assumed power as a minority government in 2006, Harper's
Conservatives had managed to alienate most of Quebec. In
Papineau itself, Tory support was abysmally low.

My political style began to be profoundly influenced
by Sophie, who, as well as having a deep, intuitive understanding of Quebec, also kept a close watch on my campaign materials and my media appearances. Whenever she
saw me veering toward anything approaching a negative
style, she was quick to tell me. She also made it clear that
she would not stand by and watch the petty feuds and fric-
tions of political life poison my personality. The reason I
got into politics, she would remind me, was to promote
my image of a better Canada, not merely to score zingers
in the next day's news cycle. Sometimes it's easy for people

who have made politics their livelihood to get caught up in the heat of battle and forget about their personal values. Sophie never does, and no matter how intense things get, she makes sure I don't either.

Still, there were times when it was necessary to speak directly about Vivian Barbot. I would emphasize that no matter how many admirable qualities Mme Barbot had as a person, the fact remained that as a Bloc Québécois MP, she represented a party that had as its goal to divide people. And that what the riding, and world, need most are politicians who are focused on bringing people together.

When I spoke at community events, I learned to keep things clear and simple when discussing Liberal policies. It wasn't because voters weren't interested in the details, but many of my interactions with voters were necessarily brief, and I needed to give the "elevator pitch" on our platform. Sophie was invaluable at times like these. Before any major event, she would ask me to go through the points I wanted to make, explaining each one in direct and simple language, and she would make me restate anything that sounded complex or confusing. It worked. Eventually I was able to explain even intricate policies, such as Stéphane Dion's ill fated Green Shift, in thirty seconds or less.

Social issues cropped up often in discussions with Papineau residents, especially among some newer immigrants who were opposed to gay marriage, abortion, and legal reform on marijuana. I could not simply pander to their position. I had to adhere to my own views, which could be a challenge when I found myself being grilled on such topics during a question-and-answer session at a mosque or church. My response would be to say, "We disagree on this, and since we are both arguing from what we regard as our core principles, there is probably little room for compromise. I hope there is enough common ground on other issues, however, for you to consider voting for me." The reaction to these words frequently surprised me. At the very least, the audience appreciated the fact that I gave them straight answers to hard questions, even if they weren't always the answers they wanted to hear.

I remember one episode early on in a Pakistani-Canadian mosque on Rue Jean-Talon when I faced a situation that would become fairly frequent: new Canadians who were very supportive of me but also very socially conservative. Alex had suggested that I avoid talking about gay marriage, but all that did was trigger my desire to talk about gay marriage.

"I know that everyone in this room supports our Charter of Rights," I told the crowd. "It's the document that forms the basis of the rights we all enjoy, including the free practice of religion. But guess what? Those rights that protect you also give gays the right to marry and give your daughter the right to marry a non-Muslim. The Charter of Rights protects freedoms for everyone. You can't pick and choose the rights you want to keep and leave behind the ones you don't like." The audience was a tough one—bearded, stern older men—but they nodded and engaged favourably in a great discussion about our shared vision for our kids and our country.

Campaigning in Papineau forced me to do a lot of hard thinking about what we Canadians mean by the term *multiculturalism*. The concept is embedded in the Charter, yet it remains the subject of widespread misunderstanding. By listening only to radio call-in shows and reading only the op-ed pages of newspapers, you might think multiculturalism is a sort of free-for-all in our society, an excuse for turning a blind eye to cultural practices that we would otherwise find repugnant or even criminal. My experience in Papineau taught me that this is mostly fear-mongering. The majority of immigrants I met were aware of the prevailing cultural norms in our country, from our religious pluralism to our attitudes about equality between the sexes and our rejection of hate speech. They accepted them fully. They also realized that Canada has just one set of laws for everyone. When it comes to enforcing the Criminal Code and the principles of family law, we do not offer special treatment according to race or religion. Many immigrant families I met in Papineau brought with them lingering animosities from their country of origin, but they accepted that Canada was a place where people come to escape old-world feuds, not to nurture them.

So what does multiculturalism mean to these people— and to me? It means a presumption that society will accommodate forms of cultural expression that do not violate our society's core values. These include the right of a Jew to wear his kippa, a Sikh to wear his turban, a Muslim to wear her headscarf, or a Christian to wear a cross pendant. When I began my campaign in 2007, Jean Charest's Liberals were in power in Quebec and the Parti Québécois had yet to launch its plans for the Quebec Charter of Values, the so-called secular charter. But these scare tactics about immigrants already were a theme in public discourse. In January 2007, at the point when the question of "reasonable accommodations" was in full swing in Quebec, the small town of Hérouxville passed a resolution banning, among other things, the stoning or immolation of women. The amazing thing about their decision was the fact that the town itself had no immigrants among its residents, and had never witnessed any social strife related to minority cultural practices.

I rejected that sort of fear-mongering then, and I reject it now. In fact, I am proud to say that in 2013 I was the first federal party leader to speak out directly against the PQ's proposed secular charter. What possible purpose could be served by excluding a single mother in Papineau from the public sector workforce because she was trying to balance her commitments to faith and to her role as a breadwinner? Had Papineau residents been presented with that choice by the PQ, many would not have chosen to shed their religious apparel. Instead, they would have been pushed back from life in the public arena, precisely the opposite result to the goal we should all strive for. We need new Canadians— indeed, all Canadians—to participate in building Canada, not to abstain.

The best way to think of multiculturalism is to picture it as a sort of social contract. Under the contract, newcomers to Canada promise to abide by our laws; teach their children the skills and language fluency necessary to integrate into our society; and respect, if not immediately adopt, the social norms that govern the relationship between Canadian individuals and groups. In return, we respect aspects of their culture that may be precious to them and harmful to no one else. Forcing a nine-year-old soccer player to remove a turban, firing a daycare worker because she wears a hijab, banning a cardiologist from the operating room because he wears a kippa—these are gestures that break our side of that social contract.

Canada is perhaps the only country on earth that is strong because of our differences, not despite them. Diversity is core to who we are, to what makes us a successful country. We live it everywhere, in small towns and big cities, all over the country. It is one of our most important and unique contributions to the world. That is why I am so quick to defend minority rights, and to promote the Charter of Rights and Freedoms. I believe that our openness is at the heart of who we are as Canadians. It has made Canada the freest, and the best, place in the world to live.

CAMPAIGNING IS BOTH ART AND SCIENCE. SOME PEOPLE are naturally good at it, others not so much. A few basic skills need to be learned and practised.

I needed to learn to be more assertive in reaching out to Papineau voters. It's all well and good to be polite, but there is no point showing up to an event just to be a wallflower.

One large public fair I attended early on featured an outdoor play for children. Many people were seated directly in front of the stage. Behind them was a large open space where adults and children were milling around during the performance. This was a perfect place to connect with voters, but I held back because I didn't want to distract attention from the play. I had to be practically yanked out of my chair and pushed toward the crowd, reminded that this was a casual atmosphere and that I was there with a purpose.

No one objected to my presence or to my efforts to engage them in conversation. Indeed, over the year and a half I spent campaigning in Papineau, I never once encountered hostility when I took advantage of situations where I could introduce myself to people. As cynical as today's voters might be about politics, most people welcome the opportunity to take the measure of a politician with a handshake and a quick chat.

After months of campaigning, it's natural for a politician to believe that everyone must know who they are. You need to keep reminding yourself that many people *don't*. In fact, many Canadians can't even name their sitting MP, let alone their *would-be* MPs. And, as I discovered, having a familiar last name doesn't help as much as you might expect.

Alex Lanthier, my campaign manager, had an effective way of making his point about improving my campaigning style. One day, I was strolling through a crowd at an event, shaking hands as I went and giving everyone a quick "Hi!," "Hello!," or "Nice to meet you!" as I passed them. Alex watched me for a few minutes before pulling me aside and saying to me slowly, "Hello. I'm Alex Lanthier. And who might you be?"

I replied, "I'm Justin Trudeau."

Alex smiled. "Good. Now tell that to everyone when you meet them. Everybody. Let them hear your name. If they don't learn who you are, you'll just be some friendly oddball who randomly shook their hand."

Genuine communication with people has always been important to me, even as a novice politician. If you're going to meet voters, you have to be patient enough to spend time learning their names and asking them what's on their minds—and then listening to what they tell you. Otherwise it's just not worth it.

That said, no matter how much you reach out to people, a hard lesson to learn is that you can't persuade everyone to vote for you. I received the most negative reactions from certain older voters who, for whatever reason, retained bitter feelings toward my father. Many of them were downright angry at me for having the gall to knock on their door. There was no way of making headway with them, no matter how long I tried looking for issues we might be able to agree on.

This was the flip side of the criticism I faced that I would try to ride the coattails of the Trudeau name. When a former politician's very name inspires red-faced indignation in a voter, you can't expect the voter to feel well inclined toward the politician's son. Whenever I would encounter these situations, I would say, "I'm pleased to meet you, even if you do not think you will vote for me," and move on to the next door.

Finally, when you're campaigning as a candidate for a major political party, it's important to remember that every syllable you speak in public will be scrutinized for mistakes, inconsistencies, political incorrectness, and ideological heresies. No matter how good your intentions may be, your opponents will be ruthless in searching for any snippet they can take out of context to discredit you. For instance, the Conservatives' first attack ad against me, in 2013, featured a completely out-of-context quote from an interview I had done about my father back in the 1990s. The only thing you can do in the face of such attacks is have faith that Canadians are smart enough to see them for what they are, and will separate the wheat from the chaff when it comes time to make a choice.

That doesn't mean you shouldn't (or can't) defend yourself. Many people enter politics thinking that all they have to do is be respectful of others and speak from the heart. But that is not enough. It is never enough. Every sentence you utter will be misinterpreted when stripped from its context and posted on Twitter.

But you can also get yourself in trouble needlessly, by making it too easy for your critics. This often happens when I try to be too clever or witty. During my outreach campaign in Papineau, a student asked a question on my website: "If an extraterrestrial came to Canada and became a citizen, would he or she—or it—be protected by the Charter of Rights and Freedoms?" I thought it was a great question, partly because it tapped into my love of science fiction, but mostly because I appreciated the thought and imagination that the kid had put into it. Ignoring advice not to answer the question, I wrote a fairly detailed tongue-in-cheek reply that underscored our commitment to diversity by saying, in effect, yes, any extraterrestrial who became a Canadian citizen could claim protection under the Charter. (For what it's worth, a past president of the Ontario Bar Association blogged that my response "was legally quite correct" and "does raise the interesting issue of civil rights for non humans.")

A few days later, *La Presse* featured a cartoon showing me telling E.T. that he had Charter rights while E.T., looking oddly like Stéphane Dion, made an obscene gesture with his famous finger!

AS MUCH AS I WENT DOOR-TO-DOOR AND APPEARED AT various events throughout the riding, my primary focus was getting to know the different organizations and community groups that served Papineau residents. It was a natural choice for me. My earlier involvement with civic action and volunteerism came through Katimavik, avalanche safety awareness, sexual-assault prevention, and other campaigns promoting specific issues and causes. Papineau was home to an assortment of grassroots groups whose different mandates touched on virtually every aspect of people's lives, especially new immigrants who lacked the means or connections to secure proper housing, child care, and job opportunities.

La Maison de Quartier Villeray, located in the heart of the Villeray neighbourhood, is one such organization. Its mission is to help people who are economically and socially isolated gain the skills and confidence they need to become productive members of the community. Volunteers from La Maison de Quartier bring meals to local residents, help people get to medical appointments, hold educational workshops for parents, and generally reach out to residents who risk falling through society's cracks. Relatively well-off Canadians who don't need help putting food on the table or getting to the doctor are often unfamiliar with organizations like this, but in Villeray and similar places all over Canada, they are the glue that holds the community together.

I have grown more and more familiar with many of these groups and their work. One of the biggest problems they have is that the staff has to spend an inordinate amount of time fundraising. These people got into the social-work field because they wanted to *help* people, not continually ask for money in order to keep operating.

The funding model for such groups typically requires that they reapply for money every year, which means the survival of these community organizations relies on the shifting priorities of politicians. Whenever these groups need money, they often have to develop some new program that resonates with the local MP, provincial legislator, or city councillor. These groups are forced to reinvent the wheel every twelve months. I told these groups that as MP for Papineau, I would champion a model that would provide regular and predictable funding streams, freeing up the staff to focus on helping people.

Part of the problem is that many of the people who work at these organizations are volunteers, a word that conveys to some the impression that the work is optional. But that description is incorrect. Volunteer work has become *essential* to communities like Villeray. Much activity in this country would grind to a halt if the volunteers at various service organizations stopped showing up.

When it comes to volunteerism, I believe politicians should lead by example. In 2008, when a volunteer group called Coalition des Amis du Parc Jarry staged their annual spring cleanup of the biggest park in the riding, I arrived in jeans and a T-shirt to do my part, along with a few of my volunteers. The borough mayor was there, along with the MP and some city councillors, all dressed in suits. We put on our gloves, picked up some shovels, listened to a speech delivered to the volunteers, and posed for a group picture. When the photo op was over, the organizers turned to the elected officials and said, "Thank you for coming." Mme Barbot, the mayor, and the councillors took this as their signal to head for their cars. But to everyone's amazement, my team headed into the park and spent the next three hours helping the volunteers clean up.

The following year, when the date for the annual cleanup arrived, all the politicians showed up in jeans and T-shirts, prepared to work instead of just posing for pictures. Our example proved contagious.

I'm convinced that we need to go beyond the established view of volunteerism and adopt something we might call *committed* volunteerism: volunteer activities managed by organizations staffed and supported through a combination of private donations and long-term government funding commitments. This was the position I took at Katimavik, and I am convinced the same model can be extended to a wide range of other groups.

Papineau also taught me a lot about the problem of income and wealth inequality in Western society. At its western perimeter, the riding of Papineau borders Mount Royal and Outremont, two federal ridings that include some of the country's wealthiest neighbourhoods. La Maison de Quartier Villeray is just a ten-minute drive from the sprawling multi-bedroom homes of Hampstead and Outremont, but where Papineau's community needs are concerned, they might as well be on different planets.

In Papineau I represent too many parents who are so poor they routinely send their children to school without breakfast. Some children barely old enough to walk to school themselves escort younger siblings to kindergarten, because their parents work shifts that prevent them from being home when their children leave for school. Many people working in local food banks are one paycheque away from needing free food themselves, people with virtually no equity and no retirement plan.

Canada's rich and poor seldom interact. This is the self-reinforcing dynamic of income inequality. A generation ago it was more common for doctors and lawyers, bricklayers and teachers, shop owners and shop workers to live in the same neighbourhood. The size of their homes and cars might be different, but they tended to shop at the same stores, stroll through the same parks, and attend the same churches. All of this gave policy-makers and legislators an opportunity to understand the problems faced by the middle class and the poor, because in many cases these people were their friends and neighbours. This is no longer the case in many Canadian cities. In some areas of Papineau you can walk for blocks without meeting someone who graduated from university or earns a six-figure income; in Outremont and Westmount it's a challenge to find a homeowner who doesn't have at least one college degree.

In some cases, the adage about being born on the wrong side of the tracks is more literal than metaphorical. Parc-Extension, the hardscrabble western neighbourhood of Papineau, is enclosed on three sides by a highway and two railway tracks. The fourth side, to the west, is marked by a chain-link fence. A few feet beyond the fence is the leafy and prosperous Town of Mount Royal. The fence is considered a necessity for residents of the Town of Mount Royal and despised in Parc-Ex, where it symbolizes a widening economic rift in our society.

When discussing problems of wealth and income inequality with friends who share my privileged upbringing, I sometimes feel like taking them to my riding and showing them first-hand the challenges many of my constituents face. Serious inequality isn't a myth, as some conservative commentators claim, nor is it a slogan for promoting class warfare. It is a stark reality sitting in plain sight for anyone who is willing to observe it.

Inequality is corrosive over time. It reinforces itself in hundreds of unseen and sometimes unconscious ways. Unless you have to face it, in communities like Saint-Michel and Parc-Ex, it is too easy to pretend that it isn't there. We need to become open to shared prosperity just as we are open to diversity. Our response to inequality, to the problems that ordinary people are having all over this country, will go a long way toward determining our success as a country. I learn anew every day on the streets of Papineau that we need to do more—a lot more—to make sure that all Canadians have a real and fair chance to succeed.

FROM LATE 2007 THROUGH THE FIRST HALF OF 2008, I continued to campaign enthusiastically in Papineau, even as the Liberal Party was enduring serious stress and strain. Stéphane Dion was an earnest, intelligent, and well-meaning leader, but when he assumed the leadership he entered a shark tank, surrounded by people still loyal to other factions and potential leaders. A more ruthless leader might have fired everyone connected with the contenders for his job and brought in his own people. But that wasn't Mr. Dion's style, and in any case, I don't know that it would have saved him.

Jean Chrétien had won the last of his three majority governments in 2000. Under Paul Martin's leadership in the 2004 election, the Liberals were reduced to minority status, and in 2006 they found themselves out of power for the first time since Kim Campbell's brief tenure in 1993. We were facing serious problems, yet some Liberal Party faithful assumed that the 2006 result had been merely a weird glitch, and the country would come to its senses in time for the next election. It had yet to sink in with them that the sponsorship scandal had alienated many voters, especially in my home province. Added to this this were hangovers from the lengthy feud between Jean Chrétien and Paul Martin, the party's lazy approach to grassroots outreach, its neglect of youth and ethnic voters, and its general sense of hubris and entitlement. It was clear to me and to many other Canadians that the Liberal Party had forgotten much that it once knew about the hard work required to earn and keep people's trust.

On September 7, 2008, Prime Minister Stephen Harper visited the governor general's official residence and asked Michaëlle Jean to call a general election, launching the busiest thirty-seven days of my life. "Kiss your wife goodbye,"

I was told. "You're not going to see her much for the next five weeks."

My days began at seven, when I stood outside one of the nine metro stations in the riding handing leaflets to commuters rushing to catch their morning trains. When morning rush hour ended I began mainstreeting at stores and restaurants throughout the riding. Many businesses were empty when I walked in, but it didn't matter. I would spend time talking with the owner and cashiers, important people who might be persuaded to put up my election signs and perhaps speak well of me to their customers.

I tried to speak with as many groups as possible, including, for example, the Association Professionnelle des Chauffeurs de Taxi du Québec, unsurprisingly an influential group of people. (How many times have you found yourself talking politics with your taxi driver?) Lunch was usually shared with volunteers in the campaign office to help keep them motivated. Afternoons were a good time to visit seniors' community centres before heading back to a metro station, greeting passengers returning home at the end of their work day. Evenings were filled making telephone calls to community leaders, encouraging them to attend the next day's events. After a few hours' sleep, I would get up to do it all over again.

It was hard work, but I loved every minute of it: the routine, the discipline, the learning. But most of all, the interaction with the people of Papineau. So much of politics is fleeting and ephemeral. And much of it is, well, *merde*.

The connections you make with the people who invest their hope and trust in you, that's what gets you through all of the rest. That's what makes it worth doing.

The results on election night produced a wide range of emotions. I had experienced pure joy when I won the Liberal nomination eighteen months earlier, and I was elated to win Papineau by a narrow margin, tallying 17,724 votes to Vivian Barbot's 16,535. But I, like other Liberals, wasn't in a particularly celebratory mood. The party as a whole got walloped, winning just 26 percent of the popular vote nationally while the Conservatives increased their minority government representation from 127 to 143 seats. I gave an upbeat speech to thank my volunteers for all their efforts, but the overall result was disappointing.

I had been so focused on my community that the contrast between the national and local results was surprising and jarring. I was one of only a couple of new Liberal MPs to gain a seat we hadn't held at dissolution. Almost immediately, I was pulled out of the real, community-level work I had enjoyed so much and into the meaningless intrigue of political brinkmanship.

That night, the prominent Quebec news anchor Bernard Derome interviewed me as part of Radio-Canada's election-night coverage. I appeared by video feed from my campaign headquarters wearing an audio headset, with a room full of joyous volunteers in the background. After congratulating me, Derome got to the juicy question that many were asking: Should Mr. Dion stay on as party leader?

The arithmetic was awkward: the party's decline from 95 to 77 seats was disappointing, but not so disastrous that it made Mr. Dion's fate obvious. Many Liberals, including me, believed he had been dealt a bad hand, and that it would have been impossible for any leader to overcome in one election cycle the serious structural problems that had built up within the party.

"Mr. Dion is a man of intelligence and integrity," I answered Derome, "who has a deep and wise vision for this country, and I have a lot to learn from him."

Derome countered by saying, "So you're telling me that you're going to defend his leadership and you're willing to give him a second chance?"

"We're not talking about leadership," I said. "The Liberal Party has a leader, and I'm very content to serve him."

"Oh, I see that you've learned your trade well," Derome scoffed, "because that's not a very clear response."

"So give me a clear question," I said.

Derome laughed. "You remind me of your father!" he said, then added, "Should Mr. Dion remain leader of this party?"

I told him that Stéphane Dion should remain Liberal leader.

"Good. Well, that's clear. Bravo!" Derome said with a flourish before signing off.

It had been a somewhat tense exchange. Reporters love stories about party infighting, and my party had been only too happy, for too long, to oblige them. Nothing would have pleased them more than to hear young Trudeau take a shot at the leader of a diminished Liberal caucus. I refused to do that. Mr. Dion is an extremely smart, decent man who had served the party—and Canada—in critical roles for well over a decade. He deserved better than to have his future influenced by idle chatter on a disappointing election night.

Besides, I had other things to focus on. I had spent almost two years on the streets of Papineau, convincing people that I was in this for the right reasons. That I was in this for them. I now couldn't wait to get to Ottawa to represent the people whose trust I'd worked so hard to earn. I hoped, perhaps naively, that the success of this grassroots approach to politics would serve as a modest example for my party.

My most strongly held hope for the Liberal Party on election night 2008 was that our defeat would teach us a valuable lesson: our connection to Canadians had grown very weak and needed to be rebuilt from the ground up, through hard work

CHAPTER SEVEN

Life as a Rookie MP

—

I N FEBRUARY OF 2007, AS I WAS PREPARING FOR THE
nomination run in Papineau, Sophie sat me down and
said: "What you are doing this year in Papineau is
going to change our life. But here's something that's going
to change it even more."

She showed me a home pregnancy test (she later admitted it was the fifth she had taken that afternoon), clearly
displaying a little blue plus sign. I was ecstatic. All my life
I had wanted more than anything to become a dad. I was
inspired by the extraordinary father I'd had, the example
he set for me to follow. Sophie and I had wanted kids
almost since the day we got married. I felt pure excitement
and anticipation.

Everything was coming together. I'd found in Sophie the partner with whom I would share my life. I'd found my calling in public service through politics. Now we were starting a family, which would serve as the reason and the motivation for everything else.

Xavier James Trudeau was born October 18, 2007, the day my dad would have turned eighty-eight. His middle name was a nod to my Grampa Sinclair. He was a fat, happy baby with his mother's green eyes and open-hearted disposition. He would grow strong and athletic, fearless on the field and in the water, but shy in new situations with new people.

Sixteen months later, on the fifth of February, 2009, Ella-Grace Margaret Trudeau followed suit. Ella was luminous and peaceful, but with a will, determination, and quick mind that soon had her running circles around the rest of the household. And her father? Well, no big surprise: she has me wrapped around her little finger.

When Xavier was born, I was a Liberal candidate for an election that was almost a year away. I took a few weeks off, and for the first few months I was home helping out almost the whole time. When Ella-Grace arrived, I was a rookie MP. She was born on a Thursday, we came home from the hospital on Saturday, and on Tuesday afternoon, I was driving back to Ottawa for votes, my paternity leave having lasted all of four and a half days, two of them on a weekend.

I had sought Sophie's blessing before running for elected office because I knew from my childhood how tough politics can be on families and relationships. But it was a different matter living it as a father and husband.

A normal week meant setting off from Montreal Monday morning for four days of Parliament in Ottawa, and returning to Montreal Thursday evening, usually for a riding event. Friday I'd spend in the constituency office in meetings and consultations. Saturday was filled with events in the riding, and Sunday was mostly, but not always, reserved for family.

The first year I was an MP, Sophie and I had rented a furnished two-bedroom apartment a fifteen-minute walk from Parliament Hill, thinking that some weeks the family would come to Ottawa together. That ended up happening not at all, since my workdays were long and unpredictable because of votes, and Sophie's entire support system, family and friends, were in Montreal. My second year, I switched to just staying in hotels when in Ottawa.

Being away from my wife and young family so much was very tough, but I also found it useful in giving me perspective. Every week, on the drive home from Ottawa, I'd ask myself a few simple questions: Was the time I spent away from my family worth it? Was I building a better future for them, focused on serving the world they would grow up in, or was I just playing the game of politics, scoring cheap points, trying to win? I can't pretend I was always able to answer the questions the right way, but just the habit of asking them was a way to keep foremost in my mind what really mattered.

To be honest, the answers came more easily than you might think. Parliament is filled with good people. They're focused on serving Canadians well, on tackling difficult issues, on trying to figure out the best ways forward for our country. Very little of that makes it into the headlines or the nightly news. My goal for my first year was to figure out a simple trick: keep my head down while holding my head high.

Alex Lanthier, who was by this time running my Hill office, helped me immeasurably. Backbench parliamentarian office budgets don't go very far, allowing for just one full- and one part-time staff member in Ottawa and two full- and one part-time in the constituency office. Alex's years of experience in Liberal ministers' offices helped us punch above our weight. He drew on some semi-retired old hands to help set things up right, and brought in bright, hard-working young people who rapidly learned the ropes.

My focus was Papineau. Among the many things I had to learn, representing the people of my constituency well was by far the most important. Not only was that my primary role as an MP, but I knew that the diverse challenges that people in my riding faced were representative of the challenges many Canadians were struggling with across the country. So serving my constituents happened on two levels: directly, through my Papineau office, on issues such as immigration and visa requests, employment insurance problems, pension delays, and other problems that people needed their federal representative's help to solve. Then, while in Ottawa, I did my best to see the legislation and policy through the lens of their likely ramifications for the people who lived in my riding.

I am sometimes asked how being a teacher has helped me to be a good parliamentarian. A good teacher is not one who has all the answers and gives them to the students. A good teacher is one who understands the needs of his or her students and creates the conditions for them to find the answers for themselves. The goal is to help them through moments of difficulty, while remaining focused on empowering them to become successful on their own. Similarly, I see the goal of a good MP as helping government create a framework for a society in which people will become engaged, successful citizens, while offering a little extra help and support to those who need it.

And in my riding, those needs required an active, strong presence from the MP. When I walked the streets in Papineau, people would approach me with all sorts of problems, some of which had nothing to do with federal jurisdiction. I heard about garbagemen who woke up a baby by making too much noise when they collected the trash, or neighbours who played music too loudly or whose cooking spread aromas throughout the apartment building. I heard about street-parking rules that got someone a fifty-dollar ticket they couldn't afford to pay as often as I heard from people confused about eligibility requirements for programs such as employment insurance and old age security. I tried, but obviously I couldn't help all these people directly.

When I couldn't, I would always take care to connect them with someone who could offer assistance.

The issue that continues to bring the greatest number of constituent inquiries in Papineau is citizenship and immigration. Anyone who cares to educate themselves about the real-life implications of Canadian policy in this area should talk to the people who visit an urban MP's riding office. Many requests are about visitor visas or family reunification immigration applications. In a typical case, an immigrant has had a baby and would like to bring a grandparent from the old country to help them with the newborn, maybe just for a few months. Our approach is to interview the couple to get a feel for the validity of their case before writing a letter of support to immigration officials.

In cases where applications for permanent residency have already had been filed, there is little an MP's office can do to advance the process. These cases can take years to wend their way through the bureaucracy, leaving applicants in limbo. The only help we can provide is to ask immigration officials for an update on the application, something private citizens often have difficulty obtaining.

The many, many encounters I've had with frustrated constituents have convinced me that the Conservatives' ballyhooed immigration reforms are much more admired in Ottawa than they are effective on the ground. We have begun to lose something vital about the country through Mr. Harper's approach. Since the early 1890s, when Wilfrid Laurier implemented the most ambitious immigration

expansion the country has ever seen, we have always understood that immigration is essentially an economic policy.
The argument that this is a conservative innovation is frequently made by those don't know the country's history
very well. The economic value of immigration has always
been recognized. We wouldn't have much growth without
it. However, people are not simply widget makers, and I
think the current policy has lost sight of immigration's most
critical role for Canada: it is a nation-building tool. From the
short-sighted restriction of the family reunification policy
to the mismanagement of the Temporary Foreign Worker
Program, we are eroding the unique Canadian insight that
people come from abroad to find a new life, not just a better
job. We should see the newly arrived as community builders
and potential citizens, not just as employees.

All of this diversity presents some interesting challenges for a newly minted MP in an urban riding, and some
of the biggest involve the mistaken assumptions many constituents have about the power Canadian MPs can wield
on their behalf. In many developing countries, politicians
are able to alter the system whenever they feel like it. They
can erase someone's tax bill or get a loved one out of jail or
force immediate action on an immigration case, all with a
single telephone call. The major problem in these countries
is getting an appointment with the local politician. Because
your problem is solved soon after you walk into their office,
accessing those politicians often takes powerful friends,
money, or both.

This explains why immigrants from some countries were often amazed that they could walk into my office on almost any Friday and speak directly to me. Politicians in their home countries were often surrounded by bodyguards. Constituents would sometimes arrive in my office with their priest or imam or a retinue of prominent community leaders to vouch for them, obviously an unnecessary step in Canada.

Interacting with constituents over the years has shown me what government looks like to the average citizen. The biggest complaint Canadians have with government is that they find its dealings with individuals to be impersonal and bureaucratic. To a certain extent, that's inevitable. A federal government addressing the needs of 35 million people must rely on computerized processes, form letters, and touch-tone phone menus. But there should also be a role for direct personal contact between citizen and civil servant. Whenever we couldn't help a constituent, she or he at least appreciated that *someone* in government was willing to discuss their situation face to face with them. Ottawa can do a better job of communicating with and assisting citizens. In fact, it *must* do better.

That's one of the reasons I took so seriously the responsibility of responding to correspondence sent to my office. Indeed, most MPs are quite diligent about that, but in my case it provided a little extra challenge. There is no need to put a stamp on a letter addressed to Parliament Hill, so Canadians will often send the same letter to their member of Parliament and then to the prime minister, to the leader of the opposition, and to any other MPs they think might be sympathetic or in a position to offer help or support. I guess it was a good sign, then, that from the very first day, my office received massive amounts of mail from across the country on an incredibly broad range of issues.

There was simply far more correspondence than we had staff for in the office on the Hill, and the Papineau office was already at full steam dealing only with riding issues, so Alex came up with a solution. We started to recruit volunteers and interns, young people, mostly students, who would each come in for a few hours a week and help respond to the piles of letters, in exchange for work experience in a parliamentary setting and an opportunity to see politics up close.

At events I'd go to, if young people came up to me and said they were interested in politics, I'd tell them they were welcome to come into my office and help out. As a result, on some days every possible spot in my small three-room office was used by young volunteers stuffing envelopes or writing on a computer, including at my own desk. I loved it: having young idealists around also helped cut through the cynicism that was far too common on the Hill.

One of the other extra challenges I had to deal with was a larger degree of media attention than my fellow rookie parliamentarians were subject to. Here, again, the work I had done on the ground in my riding over the year and a half before I was elected proved invaluable. First of all, the more I could talk about my desire to be a worthy voice for the folks in Papineau, the better I was when talking to the press. I knew the people in my riding well: their needs, their worries, their hopes and dreams. The more I spoke about them, the less I got drawn into the intrigue and speculation that reporters loved to expound upon in their "A new Trudeau on the Hill" interviews.

But more than that, hard work in Papineau helped me immeasurably in my own mindset as a parliamentarian. I remember seeing an interview with a young, newly famous actor in which he was asked about his success, and he replied that he felt so amazingly lucky to be doing what he loved that he kept wondering when someone was going to knock on his door, tell him there'd been a mistake, and take it all away. I think we've all felt that way from time to time in our lives.

But from the first day I stepped onto Parliament Hill as an MP, I have never felt that way. Not once. Given my last name, and what my opponents say about me, it might be understandable that I *could* feel that way. But I knew how hard I had worked to get elected, to gain the trust of my constituents. I had earned the right to sit in the House and no one could ever take that away from me. The tough contests I'd faced, first for the nomination, then for the election, meant that I knew, deeply, that I was where I was meant to be. And that perspective has helped me in many ways to smile and brush off the nasty negative attacks.

When it comes to the media, perhaps my biggest challenge has been learning how to scrum. Scrums are those chaotic question-and-answer sessions in which a politician in the corridors of Parliament is mobbed by reporters. Being a polite human being who was raised well, my first instinct was to actually answer the questions asked of me. Being a teacher, I would often try to explain the reasoning and justification for my answers, providing helpful examples to aid in comprehension, and making sure as I went along that I was being understood.

But a scrum is not an interview, or a speech. Reporters aren't there for detailed explanations; they are mostly looking for a pithy quote they can drop into their story or a four-second clip to use in their newscast. And the more a politician rambles on, the greater the chance that the most interesting clip won't be the most pertinent one. So the challenge is to cover the broad range of subjects reporters will ask about during a ten- to fifteen-minute scrum in a thoughtful, but concise, way.

My problem is that I actually find conversations with most journalists interesting and enjoyable. And an interesting conversation is usually one that goes off on any number of tangents, so the best journalists (like my students before them) would often try to engage me in oblique lines of inquiry. Yet the more I strayed from the core message I needed to convey, the less likely it was that that message would reach Canadians.

Scrums do deliver the odd moment of poetry once in a while. Every adult Canadian knows my father's famous line, "Just watch me." That was delivered in a scrum of sorts.

It was 1970, and he was responding to a reporter's question about how far he would go to protect Canadians from the threat of the Front de Libération du Québec. The FLQ had murdered a provincial cabinet minister and kidnapped a British diplomat. CBC reporter Tim Ralfe had been waiting, microphone in hand, for my father's car to arrive and intercepted him as he was entering the Parliament Buildings. In our current aggressively scripted era, the video of that exchange makes for amazing viewing (it's available on YouTube). There is the prime minister of Canada engaging in a spontaneous debate about a serious national security issue with an aggressive reporter whose line of questioning indicates that he believes the PM is infringing on the civil liberties of Canadian citizens. It's not just my father's direct, unequivocal, and challenging response to the question that is so surprising; it's the fact that Ralfe's question was answered at all, and in detail.

That open, free-form exchange between journalist and politician was the example I carried in my head when I arrived in Ottawa. But times have changed. In this age of Twitter, hyper-partisanship, and sound bites, few politicians—and certainly not our current prime minister—would permit themselves to be drawn into such a frank discussion with a reporter. Rather than provide direct and candid answers, today's politicians typically use reporters' questions as jumping-off points to reiterate their party's message *du jour*. There is little place for my father's scrum style in today's Ottawa. For now, at least.

ONE OF MY VERY FIRST FORMAL ACTIONS IN PARLIAMENT was to present a private member's motion on youth service. All backbench MPs get a turn, according to a random draw, to present a bill or a motion on the subject of their choosing for Parliament to vote on. A bill introduces or modifies legislation, whereas a motion usually leads to a study by committee that results in a report.

My goal was simply to get the House of Commons to take note of the importance of youth service. I had seen in my time of advocating for Katimavik how difficult it was to get parliamentarians to understand just how empowering youth service could be, not just for young people themselves but for organizations and entire communities across our country. I didn't want to score points, I didn't want to embarrass the other parties; I simply wanted to get a study done on youth volunteerism and how the government might encourage it through a framework for a national youth service. So I introduced M-299, a motion for the creation of a national policy for youth volunteer service.

When it was soundly defeated by the Conservative Party and the Bloc Québecois, I understood first hand the catch-22 faced by anyone in politics advocating for youth issues. Young people don't think that politicians care about their issues, so won't be particularly outraged when politicians vote against them. And since young people don't usually vote in large numbers, politicians don't care to invest any time or energy in issues that matter to them. Which provides further incentive for youth to disengage politically. The cycle is self-perpetuating. It would take a personal commitment of political leadership to break it.

The whole experience hardened my resolve to speak loudly and clearly for young people across the country. I would make sure that at least one strong, vocal politician was fighting for youth in Canada.

My committee work also taught me a lot about how parliamentary politics really work. A parliamentary committee is a group of MPs charged with examining legislation or conducting a study within a certain area. A proposed bill is voted on initially in the House of Commons and, if passed, sent to the relevant committee for study. Members from all parties examine it, hear from experts, witnesses, and interested parties, propose amendments and improvements, and send it back to the House for a final vote.

At least that's how it's supposed to work. In my experience, what witnesses say, or experts recommend, or opposition members propose all matters far less than the optics and the politics that surround a particular issue. In my first years I was on the environment committee, and later I served on citizenship and immigration. On the former, all the government cared about was looking as if it cared about the environment, while doing the absolute minimum it could get away with. On the latter, it felt it had all the answers already, and anyone who disagreed with or corrected it must be a rabid opposition partisan.

I remember asking probing questions of witnesses, engaging in detailed discussions of various measures or recommendations, and regularly challenging—quite success-fully, I felt—the Conservatives' smug assertions about their record of environmental stewardship. After a few exchan-ges in particular, I felt that surely I had made a modest but significant contribution either to improving the way the government would act or at least to drawing attention to the way it shamefully neglected to concern itself with the well-being of the land that sustains us all. But the truth is that, these days, most of the proceedings in committees are sword-strokes in a pond, creating only small ripples that dis-appear quickly.

EARLY IN MY FIRST TERM AS AN MP, CANADA'S POLIT-ical landscape changed dramatically. Within six weeks of the October 2008 election, the opposition parties signalled their intention to bring down the minority Conservative gov-ernment with a non-confidence vote on the government's grossly inadequate budget update. With that done we would establish a coalition government made up of Liberals with 77 seats and the New Democratic Party with their 37 seats. The Bloc Québécois agreed to support the coalition on confidence votes. But before we could take this action, the prime minister asked Governor General Michaëlle Jean to prorogue Parliament until January 2009, effectively cutting off the non-confidence vote.

In theory, the proposed coalition could have survived well into the new year. In practice, it fell apart almost immediately.

In our parliamentary system, the government is formed by whatever party or parties can gain and maintain the confidence of the House of Commons. Citizens put people in the House of Commons with their votes, and the House decides who wields power. In a majority government, one party has more than half the seats, and so becomes the government and wins all votes. But in a minority situation, no party commands the House on its own, and therefore the government is usually formed by the party with the most seats, which will then need the support of others to win votes and govern. The minority Conservatives lost that support, and the other parties were ready to come together to form a government that would have the support of a majority of the MPs in the House. This was perfectly legitimate in theory, but in practice, legitimacy also requires public support.

The Conservatives were fighting for their political survival, and they used their powerful communications machine to erode that public support, in two ways. First, they blatantly misrepresented the way in which parliamentary government actually works and what is acceptable practice within it. It was easy—and convincing—to say, "Stephen Harper won the election and now the losers want to take over the government." It was hard to counter this argument with a description of how government legitimacy is granted by the House of Commons. A sound bite beats a civics lesson every day of the week.

Second, they emphasized the fact that the coalition would require the support of the sovereigntists to govern.

Never mind that the Conservative Party itself, in minority government and in opposition, solicited and relied upon the Bloc's support to win certain votes. Now, when it suited them, they played up the easy line that "sovereigntists would control Canada." Prime Minister Harper even stated in the House that there had been no Canadian flags at the coalition signing ceremony, which was verifiably untrue.

But in politics, perception often trumps reality. We just weren't doing very well in getting out our message. The pivotal example of that came at the height of the crisis, shortly after Mr. Harper appeared on television to make his case to the public. Stéphane Dion's response to Mr. Harper's speech was supposed to air immediately afterwards, but his team could not meet the networks' deadlines. And when the final product made it to air later that evening, Dion's rebuttal appeared as a blurry, unprofessional mess that looked and sounded as though it had been recorded on a cheap cell phone. It wasn't any one person's fault, but Mr. Dion, being the party leader, was blamed. So the die was cast, and when the governor general granted Mr. Harper's request for prorogation the next day, it was game over.

The message to me and other Liberals that evening was clear. It would not be enough for us to be a party of values and ideas. We had to communicate those values and ideas with professionalism and rigour.

Stéphane Dion announced his resignation as party leader four days later, which triggered yet another leadership convention, this one scheduled for May 2009 in

Vancouver, with Michael Ignatieff, Bob Rae, and Dominic LeBlanc interested in the position. Some people had suggested that I might stand for the leadership, but there was never any intention on my part to pursue that idea. In fact, I was so uninterested in leadership squabbles that early on I assumed the neutral role of convention co-chair. Backroom manoeuvring quickly led to Dom and Bob dropping out of the race, which meant that Michael Ignatieff became leader of the Liberal Party of Canada.

Mr. Ignatieff brought a familiar blend of qualities with him. Michael was a thoughtful, charismatic, well-travelled public intellectual prepared to adapt his philosophical sensibilities for the rough-and-tumble world of politics. This profile reminded many of my father. In fact, you could make a persuasive argument that Michael had been much more accomplished before entering politics. So why did one become a very successful prime minister while the other led the Liberals to significant defeat?

Michael's lack of intuitive feel for Canadian politics—perhaps a product of his many years living outside the country—left him vulnerable. More to the point, his timing couldn't have been worse. He returned to Canada to lead the Liberal Party at perhaps its weakest moment since before Laurier's leadership in the early 1900s, and he faced opponents in the Harper Conservatives who had mastered the art of exploiting such weakness with the nastiest, most negative attack ads ever seen in Canada. When the Tories pounced on Michael, the attacks proved effective, partly because Liberals then lacked the modern fundraising abilities that would have allowed us to respond with an equal volume of retaliatory messages.

But primarily, it was because the Liberal Party had lost touch with Canadians and we had been too busy infighting to notice. We paid the price.

That said, no Liberal could have predicted just how badly we would fare in the May 2011 election. When the last ballots were counted, we were reduced to just 34 seats in the House. Stephen Harper's Conservatives finally gained their majority with 166, and the 103 seats won by Jack Layton's surging NDP knocked us to third-party status.

I was re-elected in Papineau, but election night with staff and volunteers was nonetheless funereal. We Liberals had suffered our worst defeat in the party's 144-year history. I remember thinking that it had been a long time coming and was a product of many serious errors. In a sense, I wasn't really all that surprised. I had felt in my bones that the party's connection to the country had grown perilously weak, and that this was the inevitable conclusion of a long period of disconnection and decline.

On election night, some observers seriously questioned whether the party would survive. It was not an overstatement. In just seven years we had gone from governing with a strong majority government to being in distant third place. Our leader lost his seat, and Liberals all over Canada grimly pondered the future.

CHAPTER EIGHT

The Path to Leadership

———

THE DAY AFTER THE 2011 ELECTION, THE SUR-
vival of the Liberal Party of Canada was much
more on my mind than whether I would ever
lead it. There's no way to sugar-coat it: we got trounced. It
wasn't quite as dramatic as what happened to the Progressive
Conservatives in 1993, when they went from a comfortable
majority to just two seats, but it was close.

In fact, in some ways it was worse. The PCs suffered
a sudden and calamitous shock to their system; what happened to the Liberal Party was more like the proverbial
boiling frog. Starting in the comfortable water of a majority government after the 2000 election with 172 seats, the
party was reduced to a 135-seat minority in '04 before being

sent to the opposition bench with 103 seats in '06. We then won just 77 seats in '08, my first election. Seen in this light, we couldn't write off 2011's 34-seat result as an anomaly or some sort of freak accident. It continued a long-term trend that saw the Liberal Party steadily lose almost half of its voters over the course of a single decade. I became convinced that unless something fundamental changed, our story would end just like the frog's.

Everybody had a theory about how the 2011 debacle happened. Some blamed the Conservatives' negative attack ads; others pointed fingers at their aggressive organizational and fundraising efforts; many were convinced it was the result of Michael Ignatieff's leadership. I think all these theories are too simple, and wrong. The truth was—as it often is—a lot more painful and difficult to face: Canadians gave the Liberal Party the drubbing it had earned. I know that is hard to hear for many Liberals, even now. But it is essential to remember.

Over the course of a decade in power, facing a divided opposition, the party had become focused on itself rather than on the Canadians who supported it, elected it, and had faith in it. The notion that we were Canada's natural governing party was axiomatic to many Liberals, but for me it captured perfectly everything that had gone wrong. It got to the point where it was commonplace for one Liberal or another to utter, as an article of faith, the tired old line: "The Liberal Party created Canada." As I would say when I launched my leadership campaign seventeen months later, the Liberal Party didn't create Canada; Canada created the Liberal Party.

Like too many successful organizations, the party took that success for granted and began to see it as part of the natural order of things. It forgot how it had become successful in the first place. The Liberal Party did well in the twentieth century because it was deeply connected to Canadians, in communities large and small, all across the country. It became the platform for their ideas, their hopes, and their dreams for their country. Gradually, we lost that connection. It probably started during my own father's leadership; he (to be charitable) perhaps spent less time nurturing the grassroots of the party than he might have. It culminated in the last decade when, in opposition to a minority Conservative government, too many people thought we were just one or two adjustments away from being returned to power. These were all fundamental errors. But no matter; there is little point in assigning blame. The point is that our 2011 reckoning wasn't pre-ordained. We brought it on ourselves. In the breakup between the Canadian people and the Liberal Party, the issue was with the party, not the people.

The real question I had on my mind in that spring was, Now that we've hit rock bottom, does my party get it?

As one of the best-known Liberal survivors—it's hard to use the term *winner* in the context of 2011—it was my responsibility to appear in the media during the rueful aftermath. I knew that I'd be asked whether I would seek the leadership of the party. I had no intention of doing so, and was concerned that any ambiguity on that question might trigger another round of the negative dynamic where some Liberals deluded themselves into thinking there's a shortcut back to popularity and power. My message in those interviews was extremely blunt. I said that one thing and one thing only would get us out of the hole we had dug for ourselves: hard work. I believed then—as I do now—that Canadians would judge us both by whether they felt we really got the message they had sent us, and by whether we consistently showed them the disciplined work ethic required to earn back their trust.

Like the bell that saves a staggering boxer, summer came mercifully soon after the 2011 election post-mortems. I spent most of that summer with Sophie and the kids. We went out to B.C. to recharge our batteries with family and friends. Xavier and Ella-Grace got to explore our spectacular West Coast beaches. We put the election behind us and spent a lot of time talking about our future.

That summer was a season of reflection in other ways as well. I was in my fortieth year, and wanted to commemorate it with a permanent, personal testament of sorts. When I was very young (just five or six years old), my father took us with him to Haida Gwaii, on the Pacific coast. The Haida people have lived in that truly special place, among the most beautiful on earth, for millennia. They measure their culture's history on this land on a time scale that is all but incomprehensible to Canadians who descended from settlers who came after European contact.

In a ceremony honouring my father, the Haida also conferred upon my brothers and me a privilege reserved for very few, one we of course had done nothing to earn. They made us honorary Children of the Raven. It was a touching gesture of openness, goodwill, and friendship. So as I spent that summer with my own children on the West Coast, reflecting on my future and beset by reminders of just how fleeting and transitory our own lives can be, there was something comforting about the comparative permanence of the West Coast Native presence. I thought about that kindness shown to me some three and a half decades before, and repaid it with a very modern gesture: a Haida raven tattoo, based on a Robert Davidson design, on my left shoulder. It wraps around the globe that I'd gotten years before.

I am not telling you this story to romanticize First Nations. I have spent too much time in too many remote reserve communities to be anything but clear-eyed about the challenges many First Nations, Métis, and Inuit people face. My gesture was as much about the future as it was about that past event. It is a reminder of a fundamental fact of Canadian life: we have failed to achieve a respectful, functioning relationship with First Nations. This is one of Canada's greatest unresolved questions.

In fact, I would go further than that. The predicament of First Nations, and our willingness as non-Aboriginals to abide the abject poverty and injustice that afflict so many, is a great moral stain on Canada. To take perhaps the most poignant example of our unwillingness to face these challenges head-on, there are more than 1,100 missing and murdered Aboriginal women in Canada. The government refuses to call an inquiry into the issue, and that is shameful.

That said, what troubles me most about the government's response is that the Conservatives clearly feel they exist in a political environment that will not punish them for their inaction. Nor is this simply a point about Mr. Harper. With the notable exception of Paul Martin, who through the Kelowna Accord created a framework and principles to tackle so many of these problems together, significant progress has eluded most of our prime ministers.

What progress has been made has largely come through the courts, as First Nations people litigate the Charter and other constitutional means of protecting their rights. This has to change. Canada's relationship with first peoples is definitional when it comes to our national character and is currently a practical obstacle holding our country back. The courts are telling us what we ought to have always known: First Nations communities across Canada have a right to a fair and real chance at success. They cannot be an afterthought as we develop the resources on their land.

WE WERE ON VANCOUVER ISLAND WHEN WE GOT THE tragic news that Jack Layton had died. It was impossible not to like Jack. Even though we were political opponents, I also couldn't help but admire what he had achieved in my home province of Quebec. For many observers, the Orange Wave was an overnight success, but like all such things it was years in the making. From the beginning of his leadership, Mr. Layton had made achieving a breakthrough in Quebec one of his very highest priorities. I'm sure there were many within his own closest circle who thought it was a stretch at best, but he stuck to it. He was dogged and disciplined about it; he chipped away at it over a long time until he got his opportunity. When it came, he was ready. This was one of many important lessons to be learned from his success. The fact that he was cut down by cancer so shortly after achieving that success added poignancy to the tragedy of his death. We chatted only a few times, but like almost everyone he met, I was always touched by his grace and friendliness.

Canadian public life still misses Jack Layton.

Maybe this sad news contributed to hardening my determination not to run for the leadership of my party. I'm not sure. What I am sure of is that I was one hundred percent at peace with the decision. I had convinced myself that too many Liberals would see my entering the leadership race as another potential shortcut around the monumental task ahead. I was determined to play an active role in that work, though. I told my colleagues at our first caucus meeting, in September, that I would not seek the leadership but that I was excited about the future and looking forward to rolling up my sleeves and getting dirt under my fingernails.

During this period, lots and lots of people doubted that the Liberal Party had a future. Serious authors wrote books about our imminent demise. We Quebecers were particularly aware of the party's perilous state. Outside Quebec, few Canadians appreciate the fall-out of the Gomery Commission within my home province. Whatever you might think about who did what to whom in the Liberal internal disputes, the party's basic integrity was called very much into question in the minds of millions of Quebecers. There was no wishing it away, no way to pretend it hadn't happened, no way to sidestep it with new leadership. We were in a deep hole, and the only way to win back the faith of Quebecers—and other Canadians—was hard work over time.

Despite these very real problems, I never doubted that the Liberal Party could have a future. I believe that Canadians want a truly national, non-ideological, and pragmatic party that is connected to them and focused on them. One that is focused on the hopes and dreams they have for themselves, their families, their communities, and their country. We have not always lived up to it, but at our best, the Liberal Party can be a unifying, constructive national force. One that, since Wilfrid Laurier, has tried to focus on building common ground among people whose many differences are too easily exploited as divisions by a more cynical style of politics and politician.

Among the many, many great things about democracies is that they tend to be self-correcting over time. If a government becomes too self-centred or out of touch, it gets replaced. If people want a new political movement, they will create one. I have always believed that Canadians want a party to play the central role the Liberal Party once played. The question was, in the aftermath of their worst defeat in history, could the Liberals become that party once again? In other words, could the venerable Liberal Party become the twenty-first-century movement that Canadians needed?

A couple of pivotal and auspicious things happened in the months after the 2011 election. First, Bob Rae stepped in as interim leader to provide stability and calm at the top. It's hard to overstate how important Bob's leadership was through this period. He established credibility and a professional, businesslike stewardship that the party badly needed at the time. Most important, he set an example by working hard. He wouldn't accept the prevailing wisdom of the moment, which was, overwhelmingly, that the party was about to be consigned to history's dustbin. Few people will ever appreciate how important Bob Rae's relentlessness was in ensuring that the Liberal Party didn't show up at the wake many had prepared for it in 2011.

Second, the party's base responded decisively. People across the country rose to the challenge. The much maligned, often dismissed grassroots of the Liberal Party showed up, en masse, in Ottawa for the January 2012 policy convention. I have to confess that even I was pleasantly surprised by the unmistakable enthusiasm on display everywhere that weekend. It was a lesson that reminded me of my experience in Papineau. It's easy to get caught up in what people are thinking, writing, and talking about in the Ottawa bubble when you're there. The prevailing winds can create their own atmosphere. The 2012 convention was the first time Liberals from across Canada had had the opportunity to get together since the election, and the pushback against those prevailing winds was invigorating.

In some ways, the severity of our defeat prepared the way for our rebirth. As the third-place party, we had the leeway to discuss contentious issues we would never have considered as a governing party. Interesting things began to happen. When 77 percent of the delegates at the 2012 convention voted in support of a motion to legalize marijuana, we were comfortable with the idea. We made the Liberal Party a truly liberal party when delegates decisively endorsed a motion defending a woman's right to choose whether to have an abortion.

We made equally important—if less visible—decisions to modernize the party's administration and political modus operandi. We elected Mike Crawley as party president. He had run on the slogan "A Bold, New Red," and he drove an ambitious agenda for the party to begin to professionalize fundraising. Most important, we revamped the party's structure to accommodate a new no-fee class of "supporters": Canadians who shared our values and who would be given a voice in choosing the next leader. The supporters contingent would also create a broad base for raising money and spreading the Liberal message. This helped us tap into the new networked nature of modern political movements. For decades, political parties had communicated with citizens through the broadcast media, mass mailings, and phone banks. Barack Obama's presidential victory in 2008 changed all that. Now, most important political communication is spread peer-to-peer via social media. People who set up Liberal Facebook pages or attract thousands of Twitter followers to the party may lack the time or desire to attend a party convention, but they respond with enthusiasm when they discover their place in the organizational structure.

The level of rejuvenation that took place in the Liberal Party over the years after the 2012 policy convention was outstanding, no matter how you measure it. Speaking as one very interested participant and observer, it gave me confidence that the party was finally beginning to learn its lesson and was willing to put in the hard work required to earn Canadians' trust. Young Liberals especially turned out in droves. As is often the case at critical moments, it was the young people who saw what was at stake and took the future into their own hands. Many of these committed young Canadians would become major figures in my leadership campaign, and are now growing into community leaders all across the country. Some will be candidates in the next election or in future elections. Others will apply their civic-mindedness through volunteer work. All of them are imbued with a positive attitude toward public service and the public interest. They were both the breath of fresh air and the swift kick in the pants that the Liberal Party needed, just when we needed them. That January they gave the clearest indication yet that ours was a party that would not go quietly into that good night.

Most of all, I enjoyed being present at the 2012 policy convention. It had an unmistakably positive spirit. It was free of factionalism and infighting, free of hand-wringing and finger-pointing about the past. While the 2011 disaster still stung, I got the sense that people were reflecting on it only to learn lessons to take into the future. Something important had happened to the party, and the 2012 event crystallized it. The shock of hitting rock bottom in 2011 had been the jolt that convinced Liberals everywhere that it was time to rebuild the party from the ground up.

That shared work ethic was the most hopeful sign yet that Liberals had learned the lesson Canadians had been trying to teach us for the better part of a decade: the Liberal Party has no inherent right to exist, let alone govern. We have to earn it. Serious effort was required, and the country would accept no substitutes.

We finally seemed to have a critical mass of people who got that.

As I sat in the convention centre in Ottawa listening to all those enthusiastic Canadians, young and old, from every corner of Canada, speak passionately about the country they wanted to build, I started to seriously entertain the thought that I could lead them. It's more than a little ironic, but I don't think I would have run in the end had I not ruled it out so categorically several months earlier. That intervening period gave me the peaceful detachment and serenity I needed to truly reflect on the party's prospects and think about what a better kind of liberal political movement might look like, without the necessarily self-interested distractions that planning a leadership campaign would entail. Had the fall of 2011 been filled with the "will he or won't he?" intrigue that attends potential leadership aspirants, I doubt that I would be leading the party today.

Nonetheless, at this point the idea was just beginning to germinate. I was nowhere near ready to make a definitive call. Shortly after the convention, I had a long conversation with my old McGill friend Gerry Butts, who had been a long-time principal secretary to the Ontario premier. He had left politics in 2008 to become CEO of World Wildlife Fund-Canada. I told him about the convention, and about how pleasantly surprised I was that people had showed up en masse, full of hope and ready to work. I let Gerry know that, for the first time, I was beginning to revisit my decision to rule myself out of the leadership race, and I asked him what a leadership campaign might look like. I made it clear that I hadn't made a decision; I just wanted to think through the options.

Shortly thereafter, we approached Katie Telford. I had gotten to know her when she was Gerard Kennedy's leadership campaign manager in 2006. She had subsequently worked in Stéphane Dion's office as deputy chief of staff, and I liked and trusted her. Katie is hard working, tough, honest, and wicked smart, plus had actually run a federal leadership campaign. I was glad to get her assessment of the task we had ahead of us.

Gerry and Katie, along with Daniel Gagnier, whom we recruited a few months later, are still today the core of my inner circle. Dan, a fiercely proud federalist Quebecer, may be the only person in Canadian history to have served as chief of staff to the premiers of both Quebec and Ontario. I had first encountered him when I was working on environmental initiatives in Montreal, but he had first encountered me when he was a senior civil servant helping my father repatriate the Constitution in the early 1980s.

So began about six months of quiet conversations. I firmly believe that one of the most important attributes of a strong leader is the ability to recruit excellent people to your cause. There's an old saying that fives hire threes but nines hire tens. I am convinced that with the right people on your team, you can accomplish anything. That's the approach I have taken to my leadership campaign, to candidate recruitment, to staff, and to volunteer training. Leaders too often think that the presence of strong team members indicates some personal fault or deficiency. This has led, especially in politics, to a leadership model that verges on autocracy. It's a sign of weakness and insecurity—not strength—if the best person you can enlist in your cause is the person you see in the mirror in the morning. If I earn the privilege of serving as prime minister, I want to be judged by the quality of the arms I twist, all across Canada, to actively serve our country.

All that said, sometimes you have to go with your gut, even when everyone around you thinks you're wrong. My charity boxing match with Senator Patrick Brazeau was one of those moments. Not a single one of my friends, confidants, or colleagues thought it was a good idea.

The route to the ring began in June 2011, when someone told me about an Ottawa-based white-collar amateur boxing event called Fight for the Cure. Proceeds would benefit the Ottawa Regional Cancer Foundation. Now there, I thought, was an opportunity.

The concept behind a white-collar fight is to take fit professionals and executives who are usually more used to squash or spinning, train them as amateur boxers for six months, and then pit them against each other in front of tables filled with their neighbours, friends, and clients, to whom they've sold tickets. Everyone has a great time, and a lot of money is raised for charity.

I had been training as an amateur boxer since my early twenties, and I had always liked the idea of stepping into the ring for just one real fight. I had thought that once I entered politics, that item would remain forever unchecked on my bucket list, but here was a chance to test my skills in a real boxing match, for a good cause. The bonus was that I might be able to find a hard-core Conservative to be my opponent.

When I officially signed on, in October, the party was still reeling, and hurting. I knew that Liberals could use a happy diversion of some kind to raise our collective morale, as the NDP and Conservatives were having a grand old time shellacking the once-mighty Liberals in the House. I figured if nothing else, I could provide that diversion.

The fight almost didn't take place. For all the tough talk from Conservatives, I had trouble finding a Tory willing to step into the ring. I approached a number of MPs, including Rob Anders, the brash and outspoken Calgary MP, as well as Peter MacKay, then the minister of national defence, but they declined. As I joked with Dom LeBlanc at the time, "Who knew I'd have such a hard time finding a Conservative who wants to punch me in the face!"

Finally, Patrick Brazeau, who would later get caught up in the Senate expenses scandal, took up the challenge. As anyone who knows Mr. Brazeau can attest, he is big and brawny and full of swagger. It would be a good match. I had a few inches of height on him and a longer reach, but he was much thicker around the chest and biceps. He had been trained in the Canadian Forces and held a second-degree black belt in karate. He was so physically menacing that, when the fight was announced, the question quickly became not "Who will win?" but "How many seconds will it take for Trudeau to land face down on the canvas?"

Sophie, of course, had mixed feelings about the whole thing. She knew how happy the mere idea of being in an honest-to-God boxing match made me, and she watched me revel in the gruelling training regimen. But she was genuinely worried for my safety, not least because of the nature of my opponent. I talked her through my training plan and my fight strategy, shared with her my analysis of Brazeau's strengths and weaknesses, and quelled most of her fears with a phrase I'd used before, and would use again: "Sophie, I've got this."

The fight date was set for March 31, 2012, at Ottawa's Hampton Inn, and over the next six months, I trained hard. Really hard. In Ottawa, the organizing gym, Final Round, taught all the white-collar fighters the boxing basics. But I knew the stakes would be significantly higher for me, and my opponent tougher, so I drew on a dear friend in Montreal for extra help. Ali Nestor Charles runs a mixed martial arts and boxing gym in the east end of my riding. I'd come to know and respect him through the great work he did keeping kids away from street gangs and in school. Indeed, I had spent a few hours on a couple of occasions with the kids at his gym: they'd work on finishing high school in a classroom above the ring in the morning, then in the afternoon they'd train. I know they got a kick out of having their local MP join them for both aspects of their day.

Ali is himself not just a coach and mentor but also an accomplished professional boxer. So he and I trained together regularly for those six months, and when fight night came, I was truly ready.

There is something about the purity of old-school boxing. It teaches you more than a set of technical skills. It teaches you how to remain focused despite exhaustion, and to stick with a game plan even while getting battered. Most of all, it teaches you the value of discipline and hard work. I beat Patrick Brazeau in that ring because I had a better team behind me, I had a better plan, and I had trained harder to make that plan a reality. (I'll let you draw your own conclusions about whether or not that approach applies to politics.)

A week before the fight, Matt Whitteker, my trainer in Ottawa, asked about my fight plan. I told him how I thought it would go: Brazeau would throw everything he had at me early. I'd spend the first round keeping him away with my jab and reach, and let him tire himself out. By the second round I'd have more gas than him and take the initiative, and perhaps in the third round I'd go for the knockout. Matt smiled at my confidence and teased, "Oh, you'll wait till the third round to knock him out, will you?" We both knew full well that KOs rarely happen in Olympic-style amateur boxing, and if there was one, all the smart money was on Brazeau delivering it.

But as it turned out, that's pretty much what happened. Brazeau came out in a frenzy from the start, and in the first half of the first round he landed a number of huge overhand rights that had me reeling and wondering if I'd made a terrible miscalculation with this whole thing. He hit me way harder than I'd ever been hit before, even though I'd gone up against some very tough partners during my training. But just as I was beginning to wonder how much more I could take, he stopped landing those big punches. I could hear him huffing and puffing, and suddenly I was connecting my punches and swatting away his. I ended that first round with a smile on my face, because I knew it

63. Away from Ottawa, working with volunteers in High River, Alberta, after the floods of 2013.

64. Commandeering an RV wasn't at the top of my team's favourite ideas for a summer tour in 2013, especially with Sophie eight weeks pregnant, but it turned out to be a perfect mix of showing my kids British Columbia and connecting with the communities along the way.

65. As the RV tour made its way through B.C., we made sure to stop in Nelson to show the kids Kokanee Lake, where their uncle Miche is. The lake is way off in the distance and truly is one of the most gorgeous and peaceful places on earth.

66. Every stop along the RV tour provided opportunities for great discussions with B.C.'s diverse communities, no matter how big or small. Driving the whole way enabled us to make the most of our trip through the province.

67. In good Canadian tradition, the best parties always end up in the kitchen. This job spoils me by allowing me to share in a hands-on way so many of the different cultures that are a part of the Canadian identity—though I may need to practise my roti making before heading back to this year's Diwali celebrations.

68. This isn't just a Liberal rally; it's a Liberal rally in rural southern Alberta. No matter how many votes a riding earned (or didn't earn) in the last election, I believe it is important for me to get out there and connect with people.

69. Victoria was home to one of the 800-plus-person rallies that we experienced in the West. I stayed until the very end because I was so touched by how far many had travelled across southern Vancouver Island to attend.

70. After the leadership race a year earlier, the 2014 Liberal convention in Montreal was my first chance to talk directly with our membership. The rooms may change, but speech prep always goes through the same stages—including the reminders to slow down my delivery.

71. When I'm away from home, I always try to call the kids to say good night and to check in on them in the morning. In this case, we even did it during my speech at the Montreal convention.

72. Hadrien wasn't even an hour old when his father, his grandfather, and his brother started vying to hold him.

73. In Ottawa, at a first meeting of the Economic Council of Advisors. Some people you may recognize in the room include Scott Brison, John McCallum, and Chrystia Freeland, along with many others on the phone.

74. Irwin Cotler has always been an extraordinary friend and mentor in my time on and off the Hill. Any chance to get his feedback is invaluable to me.

75. A moment of celebration with our Canadian Olympians on Parliament Hill.

76. I love nothing more than a good discussion and good challenge—leave it to a Greek mother, pictured here in Scarborough, Ontario, to call it as she sees it.

77. One of the best aspects of my job is meeting and recruiting incredible people, such as Adam Vaughan. Here we are celebrating his electoral victory in the 2014 by-election in Trinity–Spadina, Toronto.

78. As a father, I feel very lucky and blessed to be able to share experiences with my children that I once shared with my father at their age.

79. It was nice to introduce Xavier and Ella-Grace to Mr. Harper. He was very gracious.

80. Hadrien in the arms of his big sister, Ella-Grace.

81. Like my dad, I need to have fun with my children too. Hadrien and I both enjoyed this moment at the Liberal Party barbecue in British Columbia in the summer of 2014.

was already over. He'd given me his all, and I could take it, and now I was going to win. The tide fully turned during the second round, and by the third round Patrick Brazeau had had enough. He was exhausted, and the slightly panicked, slightly bewildered look in his eyes made it clear he wanted to be anywhere but in that ring. When I scored a third standing eight-count in that final round, the referee ended it. It was a TKO, or technical knockout, not a true knockout perhaps, but under Olympic rules it was the best outcome I could expect.

Only then did I look out and begin to absorb everything around me. I had been so single-mindedly focused on the fight that I had barely noticed the atmosphere. The hall was filled with Conservative MPs and ministers looking forward to seeing their guy knock a Trudeau down. The fight had been carried live on a small specialty network that heavily favours the Conservative Party. They had clearly expected a different ending. I learned that people across the country had tuned in to the fight at bars and sports restaurants after the Habs lost in a shootout to Washington. As I had hoped, the event was a morale booster for fellow Liberals, and my inbox was flooded with notes of congratulation.

I knew full well that a boxing match is a far cry from the real business of politics. No ridings were won and no policies were made that night in suburban Ottawa. But political parties are teams. They are groups of like-minded, competitive human beings. They need victories to build and maintain their spirits, especially after a string of losses.

That boxing match was the first clear victory we Liberals had enjoyed over the Conservatives in a long, long time. It felt pretty good.

As is almost always the case, lots of people were quick to read too much into that admittedly dramatic event. Some in the national media even wrote that it was the unofficial launch of my leadership campaign! The truth is, I was still a long way from making a final decision. And if I was beginning to lean more in that direction, other events were far more consequential than my knocking out a Conservative senator.

Soon after our party's successful policy convention, the NDP gathered in Toronto to plot the direction of their own party. They faced the daunting task of choosing Jack Layton's successor. Mr. Layton's tragic death had come just after he led his party to the best election result in its history. For the first time, the NDP would select the leader of the official opposition and, they fervently hoped, the next prime minister of Canada. I was much less interested in the personalities involved than in what the ultimate choice would say about the party's future.

At the time, the idea that we should merge the Liberal Party with the NDP was being discussed in some surprising quarters. Even former leaders of both parties were talking about it openly in the press. For reasons I'll go into in more detail later, I always had my doubts, but at this point I was doing all I could to keep an open mind. Some NDP leadership candidates were warm to it, most notably Nathan Cullen. He campaigned on an explicit platform of electoral co-operation with the Liberal Party. I like and respect Nathan, and I was interested to see where his campaign would go.

I have a lot of friends who have voted NDP in the past. I respect the party's history a great deal and the constructive role it has played in Canadian public life over the years. The choice the New Democrats had to make was whether they wanted to stay true to their roots or try to make the transition into the status of "government-in-waiting." In concrete terms, they had to choose whether to shed their idealism for a more conventional path to power. They seemed ruled by a desire to find a leader who seemed hard enough to stand up to Stephen Harper.

In their enthusiasm to oppose Mr. Harper and the Conservatives, I think they've been getting the big things wrong about this country. For example, Canada's prosperity depends upon our ability to develop our natural resources and get them to world markets. Every prime minister in our history would agree with that. Today, that means we have to create more environmentally sustainable ways of getting this job done, but it serves nobody to suggest that western Canada's resource wealth is a "Dutch disease" that weighs down the rest of the economy. My party learned that painful lesson under my father's leadership. Using western resource wealth to buy eastern votes is a strategy that, ultimately, impoverishes all Canadians.

In the same vein, I was wary of the NDP's willingness to play footsie with sovereigntists in my home province of Quebec. Such a strategy had ended in division and rancour for the Conservatives under Prime Minister Mulroney, not to mention near disaster for the country. The Canadian Constitution is not a plaything, nor is the Clarity Act. The first defines the common ground upon which we agree to build this country together, while the second sets out the conditions (established by a Supreme Court of Canada ruling) under which we might choose to dissolve it. These are big issues. They are fundamental questions, not to be trifled with. The NDP's blithe commitment to open the Constitution on the one hand, and to repeal the Clarity Act on the other, is a very dangerous combination. They're the kind of promises that politicians make when they don't think they'll ever have to implement them.

There's an old saying in politics that if you want to replace a government, you have to provide a choice, not an echo. As I watched their convention unfold, I couldn't help but think that New Democrats had been cowed by Mr. Harper, that they were somehow overwhelmed by his way of doing politics. It seemed to me that they had decided the only way to defeat the Conservatives was to create their mirror image, only on the left.

I could be wrong, but I don't believe that's the kind of politics Canadians want. I know it's not the kind of politics that Canada *needs*.

Which brings me to the critical point. At the thirty-thousand-foot level, I had excellent positive reasons to run for leader at some point in the future. There was my desire to serve Canada, my sense that my unique upbringing meant I had a responsibility to give back to the country. After becoming a father, I gained a deeper, more concrete appreciation for why it is so important to work hard to hand to our kids a stronger country than the one we inherited from our parents. I felt the Liberal Party was finally getting to a place where it was willing to undertake serious reform, and to do the work required to earn back the confidence of Canadians. I had two successful elections under my belt, in which I fought very difficult national headwinds, giving me confidence in my own political instincts and abilities. All these things played a role in my decision to enter the leadership race. But one more consideration was driving my thinking.

After all, this was not "at some point in the future." This was now. And in 2012 in Canadian politics, the dominant force was Stephen Harper's Conservative Party.

Many people speculated that, after finally achieving their much-coveted majority government, Mr. Harper's Conservatives would moderate their views and approach. Minority Parliaments mean the government is constantly on a campaign footing, because an election can happen any time it loses the confidence of a House of Commons in which it is outnumbered. Everything is coloured by a heightened sense of partisanship in these circumstances, the argument went, but with the comfort of his majority, Mr. Harper could afford to look longer term and plot a more thoughtful strategy for the country.

It was a nice theory, but it proved utterly wrong. If anything, the government came back more virulently partisan than ever. Their majority provided the prime minister and his people insulation from democratic accountability. Lacking the check of an empowered Parliament, their worst instincts rose to the fore. Rather than focus on the major challenges facing the country—stalled middle-class incomes, climate change, the erosion of our democracy—they seemed to focus on petty issues and on settling scores with political opponents. Worse, when they did identify major problems that needed tackling—like getting our natural resources to global markets—their "my way or the highway" approach made the problems worse instead of solving them.

In short, my conviction deepened that Mr. Harper's government was taking Canada in the wrong direction, one in which most Canadians didn't want their country to go. These Conservatives are not interested in building on the common ground where we have always solved our toughest problems. Their approach is to exploit divisions rather than bridge them. Perhaps that's an effective political strategy, but it's lousy way to govern a country, especially one as diverse as ours. Once you've divided people against one another—East against West, urban against rural, Quebec against the rest of Canada—so you can win an election, it's very hard to pull them back together again to solve our shared problems.

This is the context in which, late in the spring of 2012, I started to think much more seriously about contesting the leadership. I was leaning toward it, but I still had a way to go. My first task was to get through the remainder of the parliamentary session in Ottawa so that I could spend serious time over the summer with Sophie and the kids, making sure we were up for it. I knew from painful personal experience that politicians' families shoulder a heavy burden. Sometimes, as it was with my parents, it's too much to bear.

I had directed Gerry Butts and Katie Telford to put a team together to hash out a battle plan and to think about what a successful campaign could look like, but I always knew the decision would come down to a deeply personal, private discussion between Sophie and me. We had many long, honest talks that summer. The core strength of our marriage is that we never stop talking, and we are always open with one another, even when those conversations are difficult. I wanted to be sure she knew, from my own experience, just how rough that life can be. I recalled for Sophie that my father had once told me I should never feel compelled to run for office, "Our family has done enough," he said

My dad said this despite having never experienced the incessant, base vitriol of twenty-first-century politics. It has never been my style to engage in down-and-dirty personal tactics in any endeavour. I welcome a good tussle, and my skin is thick, but I had grown up in the reality of public life. Sophie had not, and our decision would affect our kids, in some ways, more than either of us.

With all this reflection as a backdrop, Katie and Gerry set up a three-day retreat to debate campaign strategy and plan how to execute it. I made it clear that, if I was to run for the leadership, we would run a new kind of campaign, one that engaged unprecedented numbers of Canadians. We needed to throw the doors of the Liberal Party wide open. If the party was to have a future, we would need to figure out a way to give it back to Canadians.

We got together at the end of July in Mont Tremblant. My family sat with a team of people from across the country, whom we had carefully selected for their talent, energy, and experience, to decide whether this thing could be done, and whether we could do it.

We brought together a terrific group of friends new and old. Many had extensive experience in politics, and others came from the business world and the charity sector. We had a good balance of women and men, grizzled veterans and talented newcomers, professional associates and close personal friends. Most important to me, we brought our families. Sophie was there, of course, as was my brother Sacha. But I also encouraged my team to bring their partners and kids. I knew that if we were to go through with it, and be successful, every member of the team would need their loved ones' support. Every late-night strategy session, every extended tour stop I might make, every time I had to craft a speech or talk through an issue would mean that many other people besides myself would spend hours or even days away from their families. We needed, all of us, to be aware of this fact of political life from the beginning. But the gathering would also serve to remind us that we believed political life could be reformed to respond to families with two working parents.

So, like many Canadian adventures, my leadership campaign began around a campfire.

People were on couches and in sleeping bags, tucked away in corners of the cottages we rented for the weekend. Tom Pitfield, who would go on to design the innovative data strategy we used in the campaign, had brought up a brisket of smoked meat from Montreal. We built a fire in the pit out back as people began to show up and the sun began to set. When we were all gathered, I said a few words about what I hoped we would accomplish over the course of the weekend. I talked about how it was very important that we come out of our gathering with a shared sense of purpose. I joked that if we were to win the leadership campaign, multitudes would claim to have been here at this retreat, but if we went down in flames, Sophie and I would have spent a quiet weekend alone.

In closing, I asked everyone to answer a simple but important question: "Why are you here?"

One by one, people told stories that any Canadian would recognize. Some talked about their personal histories. Navdeep Bains, a promising young Mississauga MP who lost narrowly in the 2011 election, talked about how Canada had given his family opportunities that, he worried, wouldn't be granted to the generations coming up after us. Others had specific policy issues. People talked about economic opportunity and education, natural resources and climate change, immigration and diversity. Still others had more basic reasons for being there. My Papineau riding president, Luc Cousineau, who would become the chief financial officer for my leadership campaign, said that he thought Canada was becoming a much less fair country under the Conservatives. Several Quebecers in the group expressed deep regret that their province had lost its voice in Canada's national conversation. Richard Maksymetz, a brilliant organizer who was then the B.C. finance minister's chief of staff, expressed a perspective common to the westerners in our group: that the party never really walked its talk when it came to putting that dynamic part of Canada at the heart of our political movement.

It was a heartening conversation.

When it was my turn to answer the question, I concluded very simply that I believed this country was better than its current government. Canadians are broad minded and big hearted, fair and honest, hard working, hopeful, and kind. I said Canada had some big issues to tackle, but none bigger than those we had successfully wrestled in the past. I told people that for me, the greatest of all this country's many blessings is our diversity, and that this meant the people who lead Canada need to be open-minded and generous of spirit toward all, not just toward those who agree with them and support them. Too many people were being left out and left behind in Mr. Harper's vision. I said I believed that the Conservative government's basic flaw was its smallness, its meanness, its inability to relate to or work with people who do not share its ideological predisposition. I said that Mr. Harper's extreme rigidity, his belief that disagreement and dissent are signs of weaknesses to be stamped out, would have a corrosive effect on Canadian public life over time.

In short, I said that I was there because the government needed to be replaced. I wanted to find out whether we and our party were up to doing that job.

We had many important discussions and debates that weekend, about the kind of campaign we wanted to run, the issues we wanted to promote, the problems we wanted to solve. Some of it was pretty technical; I'll spare you having to indulge my inner geek by not running on about data, GOTV (get out the vote) techniques, small-gift fundraising, and the fine points of our social media strategy. There were, however, a few sessions that deserve mentioning.

When I say we had some fundamental discussions, I mean really fundamental. The first topic of the weekend will illustrate this point. It was, Should the Liberal Party continue to exist? Should we join forces with the NDP to form a united alternative to the Conservatives? Or maybe we ought to form a new political party entirely, one that was centrist in its viewpoint and values but free of the legacy infrastructure and baggage the Liberal designation carried with it.

I'm sure it will alarm some partisan Liberals to read this, but the debate was serious. After all, it was undertaken at a time when the party had been consigned to third place for the first time in history. The patient had been stabilized under Bob Rae's exemplary interim leadership, but was nowhere near out of critical condition. Lots of Canadians still called themselves liberal, but fewer and fewer were voting Liberal. We owed it to ourselves and to our country to ask the questions directly and seriously: Was the Liberal Party in the way? Did our continued existence perpetuate Conservative rule, and therefore imperil much of what our party had fought for over the years?

These were serious existential questions. In the end, the scale tipped toward a wholehearted effort to reform and rebuild the Liberal Party, for a couple of reasons.

First, there were practical concerns with forming a new party. How could we build the infrastructure in time to be a serious option in the next election? If defeating the Conservative government was an urgent priority, a new party wasn't realistic. Moreover, I'm a pragmatic realist at heart, and I knew that any party led by somebody named Trudeau would be seen as the Liberal Party no matter what we called it.

We took the merger option out for a much longer test drive. The discussion was, after all, very much alive in the public realm. Lots of thoughtful people had given it their full-out endorsement, including former prime minister Jean Chrétien and former NDP leader Ed Broadbent. These were people of substance, whom I know and respect. Their views carried weight and were worth considering seriously. That said, I always felt the merger argument was based on faulty practical premises, the greatest of which is an overly simplified approach to the basic math. As the Ontario and western members of our group pointed out, many Liberal voters would vote Conservative before they voted New Democrat, largely on economic grounds. Gerry and Katie both argued convincingly that this was, in fact, exactly what had happened in Ontario during the 2011 federal election. When our vote collapsed in the final week, voters in the Greater Toronto Area opted for the Tories because they did not trust the NDP on the economy. Other speakers pointed out that the two parties had very different cultures, and in regions like Atlantic Canada tended to be one another's main competition. Burying the hatchet would not be straightforward.

The arguments in favour all came down to expediency of one kind or another. There were those who argued that the Conservatives would do irreparable damage to the country if they were to stay in power too long, and a merger would present the best chance to beat them in the next election. Others contended that the NDP and Liberals agree on important matters of policy, and that liberally minded Canadians were growing impatient with our unwillingness to work together to defeat the Conservatives. There was something to this latter point. It appealed to my sense that the party had become too self-centred and concerned with its own success, rather than focusing on the needs of the people we expected to vote for us.

There were good arguments on both sides, but in the

end I concluded that my disagreement with the NDP over some critically important substantive matters was simply too profound for a merger ever to work, at least for me. First, I could never support their policy to repeal the Clarity Act, a move that would effectively make it easier to break up the country. This was a non-starter for me. Moreover, there were fundamental economic policy areas (trade, foreign investment, resource development) about which I thought the NDP was deeply wrong. In fact, I think that the NDP's predisposition is to be suspicious of growth and economic success, and that their policy orientation reveals this, no matter how they try to hide it rhetorically. Liberals understand that economic growth is the foundation for all else we want to achieve in areas of social policy.

I summarized my point of view by saying that we couldn't allow political strategy, a desire for power, to trump policy and principle. That is what had gotten us into trouble in the first place. Canada needs a *better* government, not just a different government. We definitively put to rest the idea of a merger.

The other major discussion we had that weekend was about what kind of leadership campaign we would run. We considered a campaign heavy on specifics and detail. The idea was that we would publish a white paper on a big policy area every thirty to ninety days during the campaign. We ultimately rejected this strategy because we felt it grated against the spirit of openness to new ideas that we were trying to inject into the party. You can't believably commit to a new, more open way of doing policy, then publish a full platform before you have had a chance to get people's input!

We decided instead to lay out major policy markers during the campaign. We would make it clear that a party led by me would be pro-growth, that it would favour free trade, practise fiscal discipline, and support foreign direct investment. We had a long discussion about how Liberals had earned credibility on the economy through the 1990s, but hadn't focused enough on it in the past decade. As in so many areas, we had taken our hard-won success for granted, and it slipped away. Specifically, we would build our economic policy on the obvious (but too often forgotten) premise that a strong economy was one that created the largest number of good-quality jobs for the largest number of people possible. We had a detailed presentation on what had happened to the Canadian middle class over the past thirty years. People's debts were growing, but their incomes weren't. We felt that nobody in Canada was speaking to the big structural changes that were happening in the economy, which were making life harder and harder for the people who were Canada's bedrock.

We had a structured discussion about Quebec and why we felt Liberal fortunes had declined so precipitously in my home province. My own view, then as now, is that we got focused on existential issues that only a narrow band of people cared about. When the sponsorship scandal dealt the party's integrity a body blow, we had nothing of substance to fall back on. In the intervening years, Quebecers had seen one scandal after another come to light, at all levels of government. Their faith in public officials generally had been badly shaken—and this was before the Charbonneau Commission started revealing its daily horrors. The way back for the Liberal Party in Quebec was to get back to its roots, to focus on meat-and-potatoes issues like people's jobs, their pensions, their kids' economic prospects. In short, I wanted to take the way we did politics in Papineau and make it the party's calling card all over the province.

This brings me to the final, and most important, decision we made that weekend about the kind of campaign we wanted to run. Our group had a lot in common. We shared values, convictions, and plenty of experience in politics and life. Many of us had young children. We also shared a misgiving about my potential candidacy. There were lots of Liberals for whom the primary positive attribute of my candidacy was nostalgia. My last name reminded them of the party's glory days, not to mention their own. There was no way I was going to run if my campaign was going to be the political equivalent of a reunion tour for an aging rock band. We could all find something more productive to do than engage in that kind of politics.

I made it clear that I wanted to run a campaign focused on the future, not the past. I wanted to build a new kind of political movement by recruiting hundreds of thousands of people into the process. We would, of course, welcome people who had been involved in the past, but the future would belong to those who could win the hearts and minds of people who would never join a traditional political party. We would build an inclusive, positive vision for the country, and have faith that Canadians would want to take part in it.

We knew there would be naysayers. We knew that for Conservatives and the right wing more generally, the very idea of a Trudeau campaign would be loathsome. They would attack us with an intensity that would make their campaigns against the last few Liberal leaders seem like a friendly spat. Their attacks would be nasty, negative, and personal. They had millions of dollars to spend, and they would spend them attacking us. They would observe no boundaries in their efforts to destroy us. I looked around the room and asked everyone another simple question: Are you up for that?

One by one, the group said yes. They were in, and they were in for the right reasons. I would make it official a couple of months later, but Sophie and I decided then and there that we were in, too. We would try to overwhelm the politics of fear and negativity with a new kind of politics. One that sought to bring people together to build on common ground, rather than divide them into camps and exploit their differences for our political gain. One that built a common vision, staked on the many things that bring us together, and keep us together, as Canadians. One that sought to foster the very best of this wonderful country.

One that was built from the ground up, on hope and hard work.

CHAPTER NINE

Hope and Hard Work

———

THE MIDDLE OF 2012 WAS A SEASON OF SPECU-lation. For an allegedly spent political force, the Liberal Party attracted a lot of interest in its leadership. Behind the scenes, Bob Rae was seriously considering a run at the job, as was Dalton McGuinty, one of his successors as Ontario premier. There were rumours that operatives were trying to recruit Mark Carney, the talented and accomplished Bank of Canada governor, while perennial names like Frank McKenna and John Manley were tossed about in the press. As well, present and former caucus colleagues Marc Garneau and Martha Hall Findlay had dropped heavy hints that they were putting together teams to vie for the job.

Soon after I made up my mind to run, I decided that I wouldn't spend a lot of time thinking about my fellow contestants, whoever they turned out to be. I know and respect all of these people, and all the others who ended up running, but there were more important issues at stake than the competitive internal dynamic of the party. I wanted instead to focus on the kind of campaign we would run, to think carefully about the agenda we wanted to promote, both in the leadership campaign and afterwards. Out of necessity, our definition of success could not end with winning the leadership of the Liberal Party as it then existed. If we were to have any reasonable chance in a general election, I knew we had to begin a substantial rebuilding effort while seeking the leadership itself. It wasn't enough to just figure out what we needed to do to win the leadership; we had to focus on what we needed to build so that the party could win the federal election in 2015. The leadership was just a step along the way.

It would be a complex effort, we knew. Most leadership campaigns are internal contests over who is going to conduct a train that is already built, you're all on, and—if you're lucky—is moving in the right direction at a decent clip. This campaign would be very different, perhaps even unique in the history of my party. We would have to draw in the passengers, build the train as we rode it, and lay most of the track at the same time. This is where the idea of hope and hard work started to take shape. We needed both a solid work plan and the positive outlook to build the numbers and momentum that would draw in the kinds of people required to get the job done.

Through years of teaching and through my campaigning in Papineau, I had learned that I have a knack for engaging people of exceptional talent who share a positive worldview, the right values, and a level of energy similar to my own, and I knew that that's what I'd do as the Liberal leader.

Like all solid plans, ours could be summarized simply: ideas and people, team and plan, hope and hard work. We wanted, more than anything else, to reground the Liberal Party as a national political force with a solid, consistent perspective on the major issues facing Canada. This would take fresh ideas and unprecedented numbers of people, in every corner of the country. Fundamentally, we knew that this campaign would decide whether the eventual winner would have a party worth leading at all.

Paradoxically, I would start by telling Liberals that our situation was more serious than they thought. The party was at a crossroads. But contrary to what members might have read, a vote for the Liberal a vote for the short term. Time and they were willing to pitch in, hard. It was a fine balance. We didn't want to dampen anyone's enthusiasm or deflate their hopefulness; we would need those in spades to shoulder through the inevitable lows that attend political life. At the same time, we had to make it absolutely clear that hope wasn't enough, and that we knew it wasn't enough. It had to be backed up and made real by a strong work ethic, and the discipline to prove every day that we were in this for the right reasons.

The Liberal Party had given Canadians too many reasons to believe that we were out of touch with their needs, let alone their hopes for their country. If we were going to win back their trust, we were going to have to earn it the old-fashioned way. In short, we had to prove that we were in it for them. So, at the age of forty, the stage of life that Victor Hugo aptly termed "the old age of youth," I set out to deliver a sobering but optimistic message to my party. Success was possible, but it was far from certain. We needed a new mission, new ideas, and new people. The first step was to refocus the party's mission where it belongs: on the needs, hopes, and dreams of ordinary Canadians. Crucially, if that message was to be something more than a slogan, we were going to need to recruit hundreds of thousands of those Canadians to our cause.

So, on October 2, 2012, on what would have been my little brother Michel's thirty-seventh birthday, in a packed room at the community centre that is Papineau's beating heart, I launched my leadership campaign with Sophie and the kids at my side. I told the crowd that I was there because I believed Canada needed new leadership and Canadians needed a new plan. The hallmark of my campaign and—if I was successful—my leadership would be a plan for economic growth that works for middle-class Canadians. I said the current government had lost touch with the things that made our country great: fairness, diversity, the commitment to leave to our children a better country than we inherited from our parents. Most of all, the nasty divisiveness that had come to characterize the Harper government was bad for Canada, and it was up to us to put an end to it. Here, in the most diverse country the world has ever known, we need leadership that proactively seeks out common ground to build upon.

From that critical perspective, our middle-class agenda is about a lot more than economics. It recognizes that the country's strength has often been reflected in Ottawa, in our best moments of political leadership, but never has it been created there. This was another lesson that Liberals had to learn anew. I said in my speech that "it is the middle class, not the political class, that unites this country." The common hopes of ordinary Canadians, be they recent immigrants living in Surrey, B.C., or tenth-generation Canadians living in Quebec City, are this country's lifeblood. Canada needs political leaders who build on that broad sense of common purpose, not ones who emphasize the few things that divide us in order to advance their own narrow purposes.

I wanted to remind Liberals that this common ground could be found all across Canada, independent of our national government and whoever might be leading it at any given time. It was out there for us to find and to build upon it a new kind of politics.

Over recent years, it had become difficult for Liberals to tell the difference between embodying values and creating them. This is what I meant when I said "the Liberal Party didn't create Canada. Canada created the Liberal Party." Historically, my party was so successful, for so long, because it was open to all Canadians, in touch with them. It was merely the vehicle for their aspirations, not the source. But with our successes, I think Liberals forgot that. That was a very large mistake, for which the party has paid a steep price.

None of this is meant to downplay the economic aspect of the middle-class issue. That is vitally important. Canada is a harmonious country largely because of the self-perpetuating dynamic of progress. People from everywhere on earth, of every conceivable cultural background, who profess every faith, have been coming to Canada for generations. They often find greater acceptance here than in their countries of origin. Crucially, they also find greater economic opportunity. This in turn makes us more welcoming of newcomers, public-spirited about the country, and better able to appreciate and accommodate the views of people we disagree with. When we feel better off by sharing common ground, we seek it out, build on it, and expand it for others.

There's nothing preordained or God-given about Canada's success, on the economic front or any other. It happened—and continues to happen—because Canadians made it happen. When shared prosperity begins to break down, short-sighted people always emerge to point out differences and take advantage of them for their own narrow interests. I am enormously proud that Quebecers recently stood up to and rejected what was perhaps the most blatant example in our country's modern history of a narrow political movement promoting division for political gain. I always had faith that we would. That said, we need to recognize that some of the seeds of the Parti Québécois's Charter of Values were sown by economic anxiety, especially in the regions outside our big cities. We need a more inclusive plan for economic growth and jobs, or we'll see more and more of such politics.

It is true that Canada has escaped, so far, the worst aspects of the middle-class decline being felt in the United States and in less diverse developed economies. Our abundant natural resources and small population all but guarantee that we will be cushioned from the worst. We have also seen, in part because of smart policies that support them, a generation of talented Canadian women enter the workforce and reach their peak earning years, which was a significant, but one-time, boost to economic growth. We should be thankful for these positive developments. We should do our best to understand and support them, but we shouldn't allow them to hide the reality of the problem. The trend is unmistakable. The median income of Canadians has barely increased since 1980. That means your average, ordinary Canadian hasn't had a real raise in thirty years. Over the same time, the economy has almost doubled in size. The struggle of the middle class in the twenty-first century is a major problem that won't be solved easily. And it won't be solved at all if we hide from it, pretend it doesn't exist, or blame it on one group of people or region of the country or sector of the economy.

Lots of people have told me this problem is too big to tackle, and I should choose smaller ones to put at the centre of our campaign. I shake my head at those in the government and on the right generally who say there's no issue here, that we're rabble-rousing and pandering for votes; or others who accept that the problem exists, but throw their hands in the air because they see it as a function of global forces that we can do nothing about here in Canada. That first argument shows how out of touch today's Conservatives have become—after almost ten years in power—with what's really happening in the lives of ordinary Canadians. The second argument reflects how unambitious they have become for the country. We have solved bigger problems with fewer resources in this country's great history. We can solve this one too, with the right plan, and the right people to implement it, in the right way. That's what building on common ground is all about.

Armed with this hopeful message, I set out to demonstrate the "hard work" side of the equation—by example. In the first week of the campaign, after launching in Quebec, I went to Alberta, B.C., Ontario, and the Maritimes. Before it was over I visited 154 different ridings and 155 different communities. Those places I could not get to, I used every imaginable contemporary technology to reach, from Skype to Google Hangouts to Twitter chats to SoapBox. My candidacy would attract attention, sure, but I knew that attention was nothing more than an open door. If Canadians didn't like what they saw on their front porch, that door would close quickly.

And, in some places, that door would open just a crack.

I CHOSE TO MAKE MY FIRST STOP IN CALGARY FOR A reason. While I wanted my campaign to be relentlessly focused on Canada's future, I also wanted Canadians to know that I wasn't afraid to confront the ghosts of my party's past. This was especially true of those ghosts closely associated with my father. The National Energy Program of more than thirty years before still looms large over the Liberals in Alberta, even more so over any Liberal named Trudeau. So I wanted to address that directly, to let people know that I recognized the negative consequences it had had. However well intentioned it might have been, the NEP ended up inciting precisely the kind of division my father had fought his whole life to bridge, in Quebec and elsewhere. I made a commitment that day in Alberta that a Liberal Party led by me would never use western resource wealth to buy eastern votes.

The NEP was a real issue, but it was an even more potent symbol. It said to a whole generation of western Canadians that when gut-check time comes around, Liberal priorities are elsewhere. It abetted our political opponents, from Brian Mulroney to Stephen Harper, in their mission to vilify us, to foment suspicion among the next generation that our party is not *for* them, in both senses of the phrase: we would not be their advocates, and they would not be welcome among us. For more than three decades, the consequence has been that even people whose predisposition is liberal on issue after issue would never think of supporting—let alone joining—the Liberal Party of Canada.

The NEP and its aftermath teaches us many lessons— none of them for the first time—but three are most salient for me. First and most obviously, resource development and the policies we create to manage it continue to be among the handful of big issues that define our success as a country. This is arguably even truer now than it was in my father's day. It's certainly the case from an economic and environmental perspective. But it's also a unity issue, a basic issue of regional fairness. Nature did not see fit to disperse valuable commodities evenly across this land. As a consequence, there are, have been, and always will be very difficult debates about resource development in Canada. These are good problems to have, if a country must have problems. There are few that wouldn't trade theirs for ours. That said, it gets to the heart of an unchanging truth about Canada: our regional diversity has always required us to keep competing demands in fair balance. When the federal government tips the scales too heavily in favour of one region over another on a big issue, the reverberations can last a lifetime.

Second, the NEP and its fall-out is a reminder that the representative nature of politics borders on tribal at a certain level. I don't mean this in a pejorative sense at all. In a diverse country where national attachments complement strong and diverse local identities, getting the balance right is vital. You can spend weeks in coffee shops in Ponoka and Wynyard and Neepawa arguing until you're blue in the face that this or that policy is good for western Canada, but if you don't have the right people willing to make the case under your banner, you won't get far. You're also, of course, less likely to develop that good policy in the first place. That's why we placed such heavy emphasis during the leadership campaign on recruiting top-quality, authentically local leaders to join the team as senior organizers and advisors, and afterwards to run for Parliament under the Liberal banner. As advanced as our communications and research capabilities have become in politics, there is still no substitute for good people who are intimately connected to their communities.

Finally, the NEP taught us a more specific and positive lesson about western Canada. The response, in the rallying cry that Preston Manning would make famous, was "The West Wants In." It says something profoundly optimistic about westerners and encouraging about Canada that the slogan wasn't "The West Wants Out." In the entrepreneurial fashion that has come to rightly typify the West, the local response to a political movement that excluded them was to create one that couldn't live without them, and to build that movement until it governed the whole country. When you take a step back and think about it, it was an awesome achievement, maybe unparalleled in our political history.

I know this will be a controversial claim, but I think the Harper Conservatives have forgotten this basic element of the Conservative Party's success.

MOST OF THE 2012–13 LEADERSHIP CAMPAIGN TOOK place over the course of a long, characteristically cold Canadian winter. I spent a lot of that season in western Canada. Among many memorable events, none is more deeply impressed in my memory than a particularly frigid evening I spent in Kamloops. It was one of those days when the sun seemed to set not long after it had risen. We were far from anywhere that could even be charitably described as Liberal territory. After a long day in the Okanagan, we drove with campaign volunteers in a minivan from Osoyoos and Kelowna north-bound up the Coquihalla. We had booked a small room at Thompson Rivers University, expecting a modest but hearty crowd of local Liberals to show up. Just before we arrived at the event, Gerry Butts, who was travelling with me on this leg of the tour, received a phone call from our key organizer on the ground. There was a problem, of the best possible kind. More than five hundred people had shown up. We were going to need a bigger room.

Many of the people in that crowd came out because of equal parts curiosity and disappointment. Curiosity about a party that they didn't know very well, with much of what they did know not being great. I made light of this during the open question-and-answer session that was a regular feature of my public events during the campaign. A student asked me what I had learned from my father about how to practise politics. I said, "When in Salmon Arm, wave with all five fingers." On a few occasions my somewhat dry sense of humour has landed me in a spot of trouble, but this time I got an unqualified laugh.

More seriously, people had shown up because they were disappointed that the party they had chosen to represent their views in Ottawa had stopped focusing on doing exactly that. The Conservative Party owes its success to the ardent devotion of its grassroots, but Mr. Harper has turned it into a vehicle for the perpetuation of his prime ministership. It is true that no government in the past few decades has done a very good job of empowering MPs or, more generally, of figuring out how to bring our parliamentary democracy into the modern era. That said, the current government has reached new depths of message control and party discipline. I know that people who remember well its Reform Party roots find that particularly galling. I'd heard that over and over during the leadership campaign, and since, but in Kamloops that night, something clicked. Off the cuff, I said something that I would say many more times over the next couple of years in western Canada: "You elected good people to be your community's voice in Ottawa. But instead, you got Stephen Harper's voice in your community."

I have never seen so many heads nod in agreement.

Canadians want to know that their votes matter. "How Parliament works" is not a topic that will keep many

Canadians up at night. What *is* important is whether their views are taken seriously by someone who, once elected, can do something about them. Whether that person will take the time and make the effort required to stay in touch with their views *after* they are elected means even more.

People feel the effects of democratic decline over time. They know when their MP is speaking from conviction rather than merely carrying a line from her leader. Mr. Harper's extreme rigidity on this front does a disservice both to his caucus and to the Canadians who have entrusted his party with their votes. I think it's the wrong approach to leadership, and that's why I have made very specific commitments to fix the problem.

There's a balance to be struck here, of course. People need to know that when they vote for a Liberal MP, he or she will support the party's platform and values. However, party discipline ought to be confined to a small number of votes: items that contradict the Charter of Rights and Freedoms, and budget and platform items.

Getting the balance right means being consistent, thoughtful, and passionate over time on the really big issues, the ones that matter most. Willingness to compromise is often a virtue in life, and in politics your ability to compromise without betraying your core values will go a long way to determining whether you'll be successful. I argued throughout the leadership campaign that too many Canadians don't know what the Liberal Party stands for. The only way to solve that problem was to make our values explicit and to live by them, even when they lead to positions that might be controversial or prove unpopular in some quarters.

I made it clear in my campaign that the Liberal Party needs to be a liberal party. By that I meant that the core values of liberalism—equality of economic opportunity and diversity of thought and belief, which I see as the building blocks of individual freedom, fairness, and social justice— ought to be the cornerstones of the Liberal Party and its policies. I said that we needed to be a party that stood up for people's right to have a real and fair chance at success, regardless whether they were born rich or poor, where they came from, or what, if any, faith they professed.

It's one thing to say these things in the abstract; it's quite another to put them into practice. For example, I think most Canadians, however they vote, agree that our diversity is one of Canada's great success stories. Personally, I think it's our greatest. As I've said before, we might be the only country in the history of the world that is strong *because* of its diversity, not in spite of it. We have managed, through hard work and generosity of spirit, to build a prosperous and harmonious society out of the most multicultural country on earth. It has been core to who we are since before our founding. It's baked into our DNA. Our instinct to look past our differences, to seek out common ground and find common cause, kept forefathers like Samuel de Champlain alive through their first winters as surely as it has helped our modern major cities become success stories that multicultural societies the world over seek to emulate today.

You can believe all these things and still not see that this value is under great stress and strain in today's Canada. Just before Christmas 2012, something happened that impressed upon me deeply how Canada's diversity needs as much support and affirmation as it ever has. I delivered a major address on the issue (included, with other select speeches, in the appendix to this book) while under ferocious attacks from the right for daring to do so. The occasion was the Reviving the Islamic Spirit convention in Toronto, an extraordinary gathering of more than twenty thousand young Muslim Canadians. They came together to talk with one another about how they could be full participants in a pluralistic, multicultural society like Canada without losing what was remarkable and unique about their religious beliefs and cultural life.

My message was straightforward. I could think of no more fundamentally Canadian discussion than that one. How to become genuine citizens of Canada without turning our backs on our communities of origin is a struggle that most Canadians have faced throughout our history. I drew on a famous example to illustrate my point. In the late nineteenth century, Catholicism and liberalism were widely considered irreconcilable systems of belief. Freedom of conscience and religious pluralism were seen as direct challenges to the authority of the church, nowhere more so than in my home province. As is the case today, the most persuasive argument that advocates of diversity had at their disposal were the facts on the ground. It came down to: Abstract ideals are fine, but in the end we have to live together, and we all don't believe the same things. We can go down the path our ancestral countries and cultures have walked (rancour, conflict, and violence) or we can try to figure out a new, more productive and generous way to live together.

The most articulate spokesperson for this point of view was a young Quebecer named Wilfrid Laurier. He was then a new MP from an upstart political party. He believed deeply that his people, as a linguistic and religious minority within a new country where the majority was English and Protestant, should set a positive example of openness and acceptance toward those who did not share their beliefs. Faced with persuasive philosophical arguments to the contrary, Laurier laid down his trump card: we are all here, and none of us are going anywhere. We believe different things. How are we to come together to build a country if we focus on what divides us rather than on our shared interests?

I believe that Laurier's logic is as powerful today as it was in his day, if not more so. The lived reality of Canadian communities, big and small, is perhaps the world's greatest evidence against those who say harmony can't come from diversity. As has so often been the case in our country's history, pragmatism wins out over a misplaced sense of cultural or ideological purity every time. That was the point I made at the convention: dogmatism, rigidity, and intolerance are antithetical to who we are as Canadians. It is as true for a young Muslim in Mississauga today as it was for that young Catholic in Quebec City in 1877. We have always built prosperity by coming together, learning from each other's distinct perspectives, but moving beyond those differences to find common ground. That is how we have worked toward a just and prosperous country.

I WOULD COME BACK TO THESE BIG, BASIC ISSUES OVER and over again: growth that works for the middle class, and fair economic opportunity for everyone; respect for and promotion of freedom and diversity; and a more democratic government that represents all of Canada. These interrelated objectives were the pillars upon which we wanted to build the campaign, a renewed Liberal Party, and a program for governing this country. They still are. The policies that will bring us closer to these objectives have been taking shape, and will continue to evolve as we approach the next election. Having articulated our goals, we are in a position to build policies to help us achieve those goals. At the end of the campaign, I would say in Ottawa that Liberals have chosen in me a leader who will "begin, spend, and end every day" thinking about how to make this country better for ordinary Canadians.

Obviously, without the people to make it happen, we wouldn't get far. Starting with a massive volunteer recruitment campaign, we set in motion what would become perhaps the largest effort in Canadian history to engage people in politics. We took great pride in the fact that our campaign was fuelled by volunteers. Near the end of the campaign, we had more than twelve thousand people working with us all over the country, the vast majority of whom had never been involved with the Liberal Party before. We designed a flat organizational structure that allowed people to plug in easily, anywhere in Canada. Our rallying cry was that we wanted to build a movement more than a party, one that privileged results over titles. It didn't matter to us whether you had been involved with the party for several decades or five minutes. Our approach was that if you got things done, you got more things to do. It was simple, straightforward, and universally understood within the campaign. And it worked.

By the time I arrived on the convention floor in Toronto in April, we had signed up more than 250,000 supporters for the Liberal Party. Some 115,000 people would vote during that week, so that by the time I was announced as the winner of the contest in Ottawa, I could claim the support of more than 100,000 Canadians.

A few people tried their best to minimize the significance of these numbers. They argued that supporters were not really attached to the party, had treated the leadership as a novelty or fad, and would disappear as soon as the race was over. Their position was that if these people weren't even willing to spend the modest amount of money ($10) required to become a member, then how committed could they actually be? From their perspective, the leadership campaign was therefore a failure.

You don't need a PhD in modern political behaviour to know how wrong-headed that view is. It grates against every contemporary trend in organizations from charities to NGOs to political parties to churches. People just don't "join" organizations the way they used to sign up for a bowling league or a glee club in the 1950s. This does not mean that people aren't committed to serving public interest causes or aren't interested in being engaged. Anyone who spends time with today's Canadians, especially young Canadians, will find a spirit of public interest that rivals any other generation's. People today are simply more demanding of the organizations they choose to affiliate with. They want more say, more involvement, and a greater number of entry points. If you were to design an organization from scratch with the sole purpose of *repelling* large numbers of ordinary Canadians, you could do no better than the stodgy rigidity of a traditional political party.

In any case, the numbers speak for themselves. When I kicked off my campaign, the Liberal Party had fewer than 30,000 members. As of July 2014, we have more than 160,000, and the number is climbing rapidly. We know from the overlap between the people who first became interested during the campaign and the new members who have joined the party since that the "supporter" effort was an enormous success. Thought up, designed, and endorsed by the grassroots of the party, the supporter class went a long way toward rejuvenating their party.

OUR HARD WORK IS JUST BEGINNING, OF COURSE. WE came out of the leadership convention with a strong, united, and energized party. We established a clear, liberal perspective on the major issues facing the country, and we built a country-wide network of volunteers who are working every day to reconnect our party to the communities we want to serve. All of that notwithstanding, I know that Canadians retain their healthy sense of skepticism toward politics and politicians. They expect us to earn their trust, day after day. I get that.

While it is likely still a year away, the contours of the next federal election are already taking shape. I believe it will be a clear choice between competing perspectives about how to build this country. The Conservative Party, after ten years in power, is running very low on fresh ideas. They are at turns in denial about or exacerbating the major problems of our time. To ordinary Canadians, who worry that their incomes aren't growing, this government says, You have never had it so good. Canadians who want to see their country take a more responsible approach to mitigating and adapting to climate change get nought but attacks and excuses, while the impacts of climate change grow more severe and the eventual necessary responses grow more costly. The inability to get strategic infrastructure started—let alone built—stands as an indictment against the Conservatives' managerial competence. The short-sighted approach to immigration has dulled this critical nation-building tool at a time when we need it most. The reckless attacks on our public institutions, from Parliament to our Supreme Court to Elections Canada, have made this country weaker, not stronger.

The central point, however, is that all these deficiencies spring from a common root cause: the autocratic, "my way or the highway" spirit that has taken hold inside the current

Conservative Party. They seem to revel in isolating enemies and defeating them, rather than reaching out and finding a larger, shared purpose. I cannot think of a leadership style more ill suited to this strong, open-minded, and kind-hearted country. Canadians respect leaders who are unafraid to disagree with them when those disagreements are genuine and expressed respectfully. One of the Harper years' most pernicious developments is a rabid form of partisanship, the idea that politics is warfare and political adversaries are to be treated as enemy combatants. In the end, we all have to come together as Canadians if we are to get anything done. As I said to my party during my first convention after they elected me as their leader: our political opponents aren't our enemies, they're our neighbours.

As I hope I have made clear throughout this book, my approach to leadership could not be more different. I am working hard to earn this country's trust. I expect no free passes and will take no shortcuts. That is, I think you will agree, how it should be. I want to be Canada's prime minister because I think I have a better idea of this country—and better ideas *for* this country—than my political opponents do, though I do not believe they are lesser Canadians, or lesser people, because we disagree. I have a strong sense of this country, where it has been, how it became great, and how it can be even better in the future. We have problems to tackle, but they are no larger than any we have solved in our shared past. And solve them we will, the way we always have: by building on common ground.

One final word. I well understand what challenges lie ahead, for me and for those I love. This will be a tough road. I draw strength from my friends, from my family, and from the experiences that shaped the man I am today. The First Nations prayer I read at Miche's memorial still guides me now, and I can think of no better final thought upon which to leave you. Thank you for accompanying me on this journey.

O Great Spirit whose voice I hear in the winds, and whose breath gives life to all the world, hear me.
   I stand before you: one of your many children,
   I am small and weak; I need your strength and wisdom.
   Let me walk in beauty, and make my eyes ever behold the red and purple sunset,
   Make my hands respect the things you have made, my ears sharp to hear your voice,
   Make me wise so that I may understand the things you have taught my people,
   Let me learn the lessons you have hidden in every leaf and rock.
   I seek strength, not to be greater than my brothers, but to fight my greatest enemy, myself.
   Make me always ready to come to you with clean hands and straight eyes,
   So when life fades, as the fading sunset, my spirit may come to you without shame.

APPENDIX

Select Speeches

——

Nomination speech for the Liberal candidacy of Papineau

Montreal, April 29, 2007

Dear Liberal friends, bonjour, kalimera sas, buon giorno. What a great day to be a Liberal!

I want to start by sincerely thanking you for being here, for allowing me to share with you the desire—the dream—that I have of representing the riding of Papineau.

I should also start by thanking my beautiful Sophie, my own family, and my extended family, people of all ages and all backgrounds who have devoted themselves to this dream for several months. Without the tireless work of this new "riding family" I would not be here before you today.

But I am here today, and it's because of your inspiration, your example, and your support. Though I have to share with you that I have another source of inspiration, as well. In the fall of 1965, the residents of Park Ex helped send Pierre Elliott Trudeau, who listed his occupation as "teacher," to the House of Commons for the first time. Times change, and riding borders change, but what you were part of forty years ago changed Canada forever. Twenty-five years ago this month, that man gave Canada one of the most evolved tools the world had ever seen in ensuring the protection and the full exercise of human rights and freedoms. And now we are all children of that Charter. Of that we are immensely proud.

So you can understand how fiercely proud I am to be able to say that your Prime Minister Trudeau was also my dad.

But it is I who stand before you today.

My name is Justin Trudeau, and I need you, the Liberals of Papineau.

You, the volunteers of our riding, like the great lady of Villeray, Lucille Girard, who brings together the young and the old every day at the Maison des Grands-Parents; like Giovanni Tortoricci, who brings friends together at their Nicolas-Tillemont club, and who even let me win a round at scopa; and like Joanna Tsoublekas, who everybody knows fights hard week in and week out for her community through the Filia Association. It's you who define the quality of life of the riding. It's you who speak to me of your daily lives and of your hopes for the future. I want to work with you, and share your challenges and your successes.

I want to congratulate Mary and Basilio for their vital commitment to the Liberal Party of Canada. Thank you both: just look how strong the Liberal Party is in this riding today. With our leader, Stéphane Dion, and the dynamism of the fierce supporters of this riding, I know that in the upcoming election we will take back Papineau.

To get us down this road, I need you.

I want to be your standard-bearer in confronting our true adversaries, the Bloc and the Conservatives.

The Bloc wants to divide and destroy our Canada. The Conservatives want to divide us on issues of social justice. They want to divide us on the environment, on Kyoto, endangering the future of our children. They want to divide us on Canada's role in the world, with positions copied from the American right.

They want to divide us . . . I want to unite us.

And just who am I? I am Justin Trudeau. I am a man with a dream for our riding, our province, and our country, and I am a man who knows how to draw us together to make it happen. I see in Canada a place where our families are strong and supported, our elders are healthy and respected, our young people are empowered and filled with hope, and our new Canadians are embraced and encouraged to join together to build the Canada this world so desperately needs us to be. To achieve this, we will all need to work together, and it starts right now, right here, this afternoon, with your votes!

Speech announcing bid for the
Liberal Party of Canada leadership

MONTREAL, OCTOBER 2, 2012

"MAKE NO SMALL DREAMS, THEY HAVE NOT THE POWER to move the soul." —Goethe

Now that'll take courage, but more than that, it'll take hard, honest work. So let me start by telling you about the folks who taught me that best, here in Papineau.

On this side of the riding, it's Park Ex. People from every nation live here. They make this neighbourhood so vibrant. On the other side of Jarry Park, Xavier and Ella-Grace's favourite park, is Villeray, one of those solidly francophone neighbourhoods that defines Montreal. Artists and intellectuals live there, but so too do many families.

In the east side of the riding, there is Saint-Michel, where you find people like my good friend Ali Nestor—a boxer—who teaches us how to fight poverty, social exclusion, and, from time to time, Conservative senators.

This community is not just remarkable for our diversity of ideas, of cultures, of beliefs. What is truly remarkable is that this diversity thrives peacefully.

Here, we trust each other and we look to the future together.

This trust that binds us together here in Papineau is the trust that binds this country together.

My friends: I love Montreal. I love Quebec. And I am in love with Canada.

I choose, with all my heart, to serve the country I love. That's why I'm so happy to announce here, tonight, my candidacy for the leadership of the Liberal Party of Canada.

So I'm here to ask for your help, because this road will be one long, Canadian highway. We will have ups and downs. Breathtaking vistas and a few boring stretches. And with winter coming, icy patches.

But we will match the size of this challenge with hard, honest work. Because hard work is what's required. Always has been.

Canada's success did not happen by accident, and it won't continue without effort. This magnificent, unlikely country was founded on a bold new premise: that people of different beliefs and backgrounds, from all corners of the world, could come together to build a better life for themselves and for their children than they ever could have alone.

This new idea that diversity is strength. Not a challenge

to be overcome or a difficulty to be tolerated. That is the heart and soul of the Canadian success story.

That, and the old-fashioned idea of progress. The idea that we owe a sacred duty to Canadians who come after us. To work hard. To build a country that offers them even more than we had. More opportunity, more choices, more success, just as our parents and grandparents did for us.

These are the values that define and unite us.

I have seen a lot of this country. And I can tell you that those values are alive and well, from coast to coast to coast.

My fellow Liberals, these values are not the property of the Liberal Party of Canada. They are not Liberal values; they are Canadian values.

I've too often heard it said in Liberal circles that the Liberal Party created Canada. This, my friends, is wrong. The Liberal Party did not create Canada. Canada created the Liberal Party. Canadians created the Liberal Party.

The great, growing, and optimistic middle class of the last century created a big-hearted, broad-minded consensus. And built a better country. For themselves, yes. But more important, for each other, and for their children.

Canadians built medicare.

Canadians built an open and dynamic economy.

Canadians welcomed newcomers from around the world into their communities and businesses.

Canadians developed an independent foreign policy, and when necessary, bled for our values in faraway lands.

Canadians brought their Constitution home.

Canadians demanded that their inalienable rights and freedoms be placed above the reach of politics.

And Canadians balanced the budget.

The Liberal Party was their vehicle of choice. It was the platform for their aspirations, not their source.

When we were at our best, we were in touch, open to our fellow citizens and confident enough in them to take their ideas and work with them to build a successful country.

If there is a lesson to be drawn from our party's past it is not where we landed but how we got there. We were deeply connected to Canadians. We made their values our values, their dreams our dreams, their fights our fights.

The time has come to write a new chapter in the history of the Liberal Party.

This will be a campaign about the future, not the past. I want to lead a movement of Canadians that seeks to build, not rebuild. To create, not recreate.

After all, we live in a very different world, my friends. Twenty years ago, I was part of the first graduating class at my university to get email. I was of the last group of pre-Google high school teachers. And now, my kids don't know there was a world before BlackBerrys.

But if the way we will build it is new, what we have to build is timeless.

We know what Canadian families want. Good jobs. A dynamic and growing economy that allows us to educate our kids as they mature, and to care for our parents as they age.

We want a compassionate society that pulls together to

help the vulnerable, and gives the less fortunate a chance at success.

We know that Canada is the freest society on earth because we trust each other. So we want a government that looks at Canadians with respect, not suspicion. That celebrates the Charter of Rights and Freedoms. That believes in your choices, your values, and your liberty.

Some say that youth carry our future. I say youth are an essential resource for our present. We need to empower all young Canadians, through world-class education, through rich and relevant work experience, and through opportunity to serve their communities and their world. Their voices, their choices, matter deeply, as do their actions: they are already leaders today.

And directly, to our First Nations, the Canadian reality has not been—and continues to not be—easy for you. We need to become a country that has the courage to own up to its mistakes and fix them together, people to people. Your place is not on the margins. It is at the very heart of who we are and what we are yet to become.

We want a foreign policy that will give us hope in the future and that will offer solutions to the world.

We want leadership that fosters and celebrates economic success in all regions of the country. Not leadership that seeds resentment between provinces.

We need to match the beauty and productivity of this great land with a new national commitment to steward it well. My generation understands that we cannot choose between a strong and prosperous economy and a healthy environment. The Conservative approach may work for a few, and for a while. But we know we can't create long-term prosperity without environmental stewardship.

We need to learn what we have forgotten. That the key to growth, to opportunity, to progress, is a thriving middle class. People with good jobs. Families who are able to cope with modern life's challenges.

A thriving middle class provides realistic hope and a ladder of opportunity for the less fortunate. A robust market for our businesses. And a sense of common interest for all.

The great economic success stories of the recent past are really stories of middle-class growth. China, India, South Korea, and Brazil, to name a few, are growing rapidly because they have added hundreds of millions of people to the global middle class.

The news on that front is not so good at home; I don't need to tell you that. You, like our fellow Canadians all over the country, live it every day. Canadian families have seen their incomes stagnate, their costs go up, and their debts explode over the past thirty years.

What's the response from the NDP? To sow regional resentment and blame the successful. The Conservative answer? Privilege one sector over others and promise that wealth will trickle down, eventually.

Both are tidy ideological answers to complex and difficult questions. The only thing they have in common is that they are both, equally, wrong.

We need to get it right. We need to open our minds to new solutions, to listen to Canadians, to trust them.

And as we face these challenges, the only ideology that must guide us is evidence. Hard, scientific facts and data. It may seem revolutionary in today's Ottawa, but instead of inventing the facts to justify the policies, we will create policy based on facts. Solutions can come from the left or the right; all that matters is that they work. That they help us live—and thrive—true to our values.

Because middle-class growth is much more than an economic imperative.

The key to Canadian unity is the shared sense of purpose so hard to define but so deeply felt. The sense that we are all in this together. That when Albertans do well, it creates opportunities for Quebecers. That when Quebecers create and innovate, it echoes across the country and around the world. That whether you're in Saint-Boniface or St. John's, Mississauga or Surrey, we have common struggles and common dreams.

It is the middle class, not the political class, that unites this country. It is the middle class that makes this country great.

We know some Quebecers want their own country. A country that reflects our values, that protects our language and our culture, that respects our identity.

My friends, I want to build a country too. A country worthy of my dreams. Of your dreams. But for me, that country reaches from the Atlantic to the Pacific, from the Great Lakes to the Grand North.

Quebecers have always chosen Canada because we know it is the land of our ancestors, who built this country from east to west. They were here to write the first chapters of the great Canadian history of courage, liberty, and hope. We have left our footsteps everywhere.

Will we put this history aside now because people of other languages came after us with the same dream of building a better country ? Of course not. Our contribution to Canada is far from over.

I want the Liberal Party to be once again the party that promotes and cherishes the francophone reality of this country. I want my party to support francophone communities across the country. And I want the Liberal Party to be once again the vehicle for Quebecers to contribute to the future of Canada.

Now, my candidacy has been the source of some speculation over the past months. The odd newspaper article has been written. Some have been very odd indeed.

But I said to Liberals after the last election that we need to get past this idea that a simple leadership change could solve our problems. I believe that still. My candidacy may shine a few extra lights upon us. It may put some people in the bleachers to watch. But what we do with that opportunity is up to us. All of us.

And when Canadians tune in, we need to prove to them that we Liberals have learned from the past, yes. But that we are one hundred percent focused on the future.

And not the future of our party: the future of our country.

I am running because I believe this country wants and needs new leadership. A vision for Canada's future grounded not in the politics of envy or mistrust. One that understands, despite all the blessings beneath our feet, that our greatest strength is above ground, in our people. All Canadians, pulling together, determined to build a better life, a better Canada.

To millions and millions of Canadians, their government has become irrelevant, remote from their daily lives, let alone their hopes and dreams. To them, Ottawa is just a place where people play politics as if it were a game open to a small group, and that appeals to an even smaller one.

They do not see themselves or their values reflected in Ottawa.

My friends, we will do better.

This is not a personal indictment of Mr. Harper or Mr. Mulcair. On the contrary, I honour their commitment and their service. But I think they are both dead wrong about this country. And, I want to tell you, together, we can prove it.

There will be many highs and lows between now and April. And if we work hard and find success, I know there will be many, many more between then and 2015.

I do not present myself as a man with all the answers. In fact, I think we've had quite enough of that kind of politics.

But I do know I have a strong sense of this country. Where we've been, where we are, and where we want to go. And I believe I can bring new forces to bear on old problems. I can convince a new generation of Canadians that their country needs them. That it values their energy, ingenuity, and vision. Together, we can convince young Canadians that serving this great country is its own reward.

I promise you this: if you entrust me with the privilege of leadership, I will work long, hard, and tirelessly. I learned first-hand from the people of Villeray, Saint-Michel, and Park Extension that there are no shortcuts, no easy ways to earn trust and support. You have to work at it, day in and day out.

Because that's what it's going to take, and that's what Canadians deserve.

Think about it for a moment: When was the last time you had a leader you actually trusted? And not just the nebulous "trust to govern competently," but actually trusted, the way you trust a friend to pick up your kids from school, or a neighbour to keep your extra front door key? Real trust? That's a respect that has to be earned, step by step.

I feel so privileged to have had the relationship I've had, all my life, with this country, with its land, and with its people.

From my first, determined steps as a toddler to my first, determined steps as a politician: we've travelled many miles together, my friends . . .

You have always been there for me. You have inspired me, and supported me in good and more difficult times. And you have made me the man and the father I have become.

I chose today to launch this campaign because it is my little brother's birthday. Michel was killed in an avalanche, doing what he loved, in the country that he loved as much as anyone I have ever known. Michel would be thirty-seven years old today. Every day, I think about him and I remember not to take anything for granted. To live my life fully. And to always be faithful to myself.

At Michel's funeral, my father read from the First Letter of Saint Paul to the Corinthians. Paul wrote, "When I was a child, I spoke as a child. But now that I am a man, I put away childish things."

It is time for us, for this generation of Canadians, to put away childish things. More, it is time for all of us to come together and get down to the very serious, very adult business of building a better country. For ourselves, for our fellow Canadians, and for our children.

We Canadians live in a blessed country. We are the most diverse people on earth, yet we are peaceful. We are tough, but we are compassionate. We are confident, but we work hard and we earn it. We have resources that are the envy of the world.

Let us pledge to one another to match those resources with resourcefulness. Let us rededicate ourselves to the glorious, improbable work-in-progress that is Canada. And to serve its people through the only party willing to speak to and for all Canadians: the Liberal Party of Canada.

So tonight, Sophie and I draw on our love for our family and offer up all we have in service to Canada, and to each and every one of you.

Join us.

Speech delivered at the Liberal Party of Canada Leadership National Showcase

Toronto, April 6, 2013

I stand before you a son of Quebec. A grandson of British Columbia. And a servant of Canada.

These Canadians you just met [in the introduction video] are a few of the thousands I've had the honour to meet, to talk with, and to learn from over the past six months.

Their stories are remarkable. Remarkable because they are no··················· in Canada.

With hope and hard work, every day Canadians live the values that unite this country. Optimism, openness, compassion, service to community, generosity of spirit.

My friends, our party must be their party.

We must convince Chanchal that we share his work ethic, his desire to serve, his optimism about the future.

We must prove to Penny that we are in it for her. That we understand the burdens she carries, every day, to make life better for her kids, her neighbours, her community.

We must build with Justine and Ali a country worthy of their dreams and show them that Canadians across our land already share the same values as Quebecers: integrity, openness, and community engagement.

To those who think that Canadians do not share common values, I encourage you to spend more time in this country. All of this country.

My fellow Liberals, my message to you is simple. To lead Canada, we must serve Canadians. And we must prove it with acts more than words. I say that not as a son who learned it from his father, but as a father who every day learns that from his kids.

The Conservatives have forgotten about the value of service. The only time they talk about "community service" these days is when it's a punishment for a crime. And anyway, the only person Mr. Harper wants his caucus to serve is their leader.

Well, that's not good enough. We need to be a party of community leaders, devoted to community service. That's why I am calling for open nominations for all Liberal candidates in every single riding in the next election.

Mr. Harper is showing us how governments grow out of touch. Canadians are getting tired of the negative, divisive politics of the Conservatives. And are disappointed that the NDP, with Mr. Mulcair, has decided that if you can't beat them, you might as well join them.

Mr. Mulcair and Mr. Harper are masters of the politics of division. They are content to exploit differences and disagreements to further their own interests.

East against West, Quebec against the rest of Canada, the wealthy against the less fortunate, cities against regions, and so on.

This is old politics. But in the short term, it can work. It was how Mr. Harper's government was elected.

We need to be better than that. We are an optimistic, hard-working, problem-solving people. Canadians want a positive alternative that brings new solutions, new ideas, and a new way of doing politics. I'm more convinced than ever that if we work hard every day between now and then, the Liberal Party of Canada will be that positive choice in 2015.

So let me be perfectly clear on one point.

I want to be your leader because I want to work with you, and with millions of Canadians, to build that positive alternative to the Conservatives. One that Canadians will choose freely because we will have earned their trust.

Canadians don't just want a different government. They want a better government.

Those who think we need to win at any cost—whatever the means—are mistaken. It is a mistake to believe that just getting rid of this government will make all of Canada's problems disappear.

This is a naive and simplistic way to approach our future.

We are facing real and significant challenges.

Middle-class Canadians have seen their incomes stall, while their costs go up and their debts explode. Simply getting rid of Mr. Harper will not get them their first real raise in thirty years.

Young Canadians will not get jobs just because Mr. Harper is gone.

Quebecers will not automatically re-engage at the heart of our federation simply because Mr. Harper is no longer prime minister.

Our international reputation on the environment will not be restored the day after Mr. Harper leaves.

The truth is, Canadians want to vote *for* something, not just *against* somebody. They want to vote for a long-term vision that embodies our values, our dreams, and our aspirations.

They will not get that vision from a Frankenstein's monster, at war with itself over fundamental issues like the Constitution, natural resources, and free trade. It would fail in its primary goal: it would extend, not end, Mr. Harper's career.

From Ponoka, Alberta, to Île-des-Chênes, Manitoba, to Edmundston, New Brunswick, Canadians are hoping that we have learned that lesson. Over the course of this campaign, I began to describe for Canadians a vision of this country that is very, very different from this government's.

Our highest economic objective will be prosperity for the middle class and those Canadians who are working hard to join it. Our grounding principle will be equality of opportunity. Our agenda will develop our skills, support our vulnerable, attract investment, and expand trade.

It is a vision that embraces diversity. One that recognizes Canada is strong because of our differences, not in spite of them. One that believes deeply in federalism, balancing national priorities with regional and local means of meeting them.

A vision that sees newcomers to this country as community- and nation-builders; as citizens, not just employees or a demographic to be mined for votes.

Ours is a vision that knows economic prosperity and environmental health can—and must—go hand in hand in the twenty-first century. We will not ignore science, or shy away from tough, urgent issues like carbon pricing. Nor will we succumb to easy politics by demonizing one sector of the economy or region of the country.

A Liberal Party led by me would never use western resources to buy eastern votes.

We will stand for national unity by offering Quebecers and all Canadians a progressive political project that rallies us all. We will be audacious and ambitious, because this country is greater than the sum of its parts

Our foreign policy will promote peace, democracy, and development. Canada must be a key player on the world stage, bringing forward positive debate and discussions—not divisive ones as is the case today.

My fellow Liberals, make no mistake about it. With me as your leader, you will get a clear, positive vision for

Canada. We've begun to lay it out in this campaign. We've focused on the big issues like the prosperity of the middle class, a healthy democracy, and a sustainable economy.

It's a vision that you and I are going to finish, together, with Canadians.

That is doing politics differently.

If we work hard and stay optimistic, we will put forward an irresistible alternative to the Conservatives thirty months from now. Irresistible not because it is Liberal, but because it will be 100 percent, undeniably Canadian.

It won't be easy. Nothing worth doing ever is. But that is the path to victory in 2015.

Hope, my friends, yes. Always hope. But more than that. Hope and hard work.

You see, the biggest problem with Mr. Harper's government is not that they're mean-spirited. It's that they're unambitious.

After all, what is the Conservatives' economic message these days? That Canadians should be happy we don't live in Europe?

What's worse, the Conservatives use our challenges as opportunities to demonize their opponents and divide Canadians, not to find solutions.

It is up to us, the Liberal Party, to say that the Conservative way of doing politics is not good enough. Canadians are better than their politics. Canada deserves far better.

Now, there are those who ask me, What makes you think you can take this on?

To them, I say this: I have lived and breathed every square kilometre of this country from the day I was born. I've lived and worked in the East and the West, in French and in English. I am proud to have lifelong friends, colleagues, and supporters from the Arctic archipelago to Point Pelee.

And I have met, talked with, and learned from more Canadians in the past six months than Mr. Harper has in the past six years.

I have been open to Canadians my entire life. And because of that, I have a strong sense of this country. Where it has been, where it is, and where Canadians want it to go.

And what is it with Conservative attacks on teachers? They've never met a teacher they wouldn't pick a fight with. I am fiercely proud to be one of the hundreds of thousands of Canadians who belong to the teaching profession. And let me tell you this, my friends, this teacher fully intends to fight back.

In closing, I want to share a story with you.

Many of you know that today marks an anniversary. Exactly forty-five years ago tonight, a gathering of Canadians made my father leader of the Liberal Party of Canada.

Many Canadians have approached me over the course of this campaign to share stories about my dad. So let me tell you a special one.

I met Constable Jeff Ling at Loyalist College in Belleville. It was at the end of a long morning. Constable Ling came up to the front of the room to give me a gift. I recognized it instantly. It was a picture of Dad and me.

You've probably seen it. I was about two years old and Dad was hurrying up to Rideau Hall, with me tucked roughly under his arm.

Both Dad and I are looking at an RCMP officer. He's dressed in full uniform and saluting us crisply.

That picture means as much to Jeff as it does to me. Because that officer was his father.

What moved me was that here Jeff was, serving his country a generation later, with the same dedication and quiet pride as his father. In that moment, he evoked the thousands of Canadians I had the unique honour of growing up with. Men and women for whom service to Canada was its own reward.

I know there are those who say this movement we're building is all about nostalgia. That it's not really about me, or you, or Canada. Let's face it: they say that it's about my father.

Well, to them I say this:

It is. It is about my dad. And Constable Ling's dad. And our mothers. And yours. It's about all of our parents and the legacy they left us. The country they built for us. Canada.

But we know now what they knew then. It's more about the future than the past. It is always, in every instance, about our children more than our parents' legacy.

That with hope and hard work, we can make progress happen. That we can leave a better country to our kids than we inherited from our parents.

Progress. That is the core value of the Liberal Party.

That is why generations of Canadians, from every corner of our land, and every walk of life, poured their heart and their soul and their ideas and their sweat into our party.

I said back in October that the Liberal Party didn't create Canada. Canada created the Liberal Party. Well, the last six months have taught me that maybe, just maybe, Canadians are willing to do that again.

We can lead the change that so many Canadians want to make happen.

I'm asking you for your time, for your smarts, for your hope, and your hard work.

And this week, I'm asking for your vote to become the next leader of the Liberal Party of Canada.

Join me, join us, and our work will make us proud. Believe, now and always, in our country.

Thank you.

Liberal Party of Canada
Leadership Acceptance speech

Ottawa, April 14, 2013

Thank you, my friends, thank you.

Normally I'd start by thanking family and friends for putting up with my absences and allowing me to go off and campaign, but that's not exactly right. My decision to seek the leadership was never in spite of my responsibility to my family, but because of it. And therefore family and friends were always at the very heart of this campaign. We did this together.

Thank you, Sophie. Thank you, Xavier and Ella-Grace.

To my fellow candidates, Joyce, Martha, Karen, Deborah, Martin, David, George, and Marc, and to the thousands of Canadians who worked on your campaigns, I want to say: we are not adversaries but allies. Your courage, intelligence and commitment will continue to bring honour to the Liberal Party of Canada.

And for the health of this party, the hard work he has done, I want to thank, from the bottom of my heart, my friend, my colleague, and a great Canadian, Bob Rae. Bob, we continue to need your leadership, your wisdom, and your unparalleled commitment to the country and to our party.

This has been a great campaign. We are fiercely proud that it has been fuelled by volunteers. More than twelve thousand Canadians stepped up. Thank you for your dedication to making this wonderful country even better.

Like every effective organization, this one has had principled, brilliant, and generous leadership: Katie Telford and Gerald Butts. My friends and compatriots. Thank you for what you've done, for what you're doing, and for what we are going to do together. Rob and Jodi, George, Aidan, and Ava, thank you for sharing Gerry and Katie with us.

My fellow Liberals, it is with great respect for those who have stood in this place before me, and great resolve to do the hard work required in front of us, that I accept, with humility, the confidence you have placed in me. Thank you. All of you. For your trust. For your hope. For choosing to be part of this movement we're building.

And on this lovely spring evening in our nation's capital, I am honored to stand with you, proud to be the Leader of the Liberal Party of Canada.

My friends, this is the last stop of this campaign. But the first stop of the next one.

Over the past six months, I have been to hundreds of communities from coast to coast to coast. I've met, talked with, and learned from thousands and thousands of Canadians. And because of your hard work, more than one hundred thousand voters have sent a clear message: Canadians want better leadership and a better government.

Canadians want to be led, not ruled. They are tired of the negative, divisive politics of Mr. Harper's Conservatives. And unimpressed that the NDP, under Mr. Mulcair, have decided that if you can't beat them, you might as well join them.

We are fed up with leaders who pit Canadians against Canadians. West against East, rich against poor, Quebec against the rest of the country, urban against rural.

Canadians are looking to us, my friends. They are giving us a chance, hopeful that the party of Wilfrid Laurier can rediscover its sunny ways. Hopeful that positive politics has a fighting chance against the steady barrage of negativity that you and I both know is coming soon to TV screens across Canada. The phone messages, our volunteers tell us, have already started.

To adapt a sentiment from the great American president Franklin D. Roosevelt: never before in this country have the forces of negativity, cynicism, and fear been so united in their hostility toward one candidate.

The Conservative Party will now do what it does. It will try to spread fear. It will sow cynicism. It will attempt to convince Canadians that we should be satisfied with what we have now.

For at the heart of their unambitious agenda is the idea that "better" is just not possible. That to hope for something more from our politics and our leaders—more humanity, more transparency, more compassion—is naive, and inevitably will lead to disappointment. And they will promote that divisive and destructive idea with passionate intensity. They will do so for a simple reason: they are afraid.

But—and I want to make this perfectly clear—my fellow Canadians, it is not my leadership that Mr. Harper and his party fear.

It's yours.

There is nothing that these Conservatives fear more than an engaged and informed Canadian citizen.

My friends, if I have learned one thing in this life, it's that our country is blessed with countless numbers of activist citizens, from all walks of life, and of all political views. They have come out by the thousands over the course of this campaign.

They've gathered by the hundreds in places like Ponoka, Alberta, and Oliver, B.C., Prince Albert, Saskatchewan, and Île-des-Chênes, Manitoba. Canadians who thought they were sending community leaders to be their voice in Ottawa, but instead got only Mr. Harper's voice back in their communities.

We've seen their hopeful faces in crowds of Canadians gathered in Windsor and Whitby, Mississauga and

Markham. Middle-class Canadians who are putting much into the economy and getting too little in return.

We've seen hard-working Atlantic Canadians from Edmundston to Halifax, from Summerside to St. John's, who have decided that this is a government that does not share their values. (To my friends in Labrador, I look forward to seeing you very soon.)

We've met young Aboriginal leaders from all across this country, from Tk'emlups to Whapmagoostui, who are simply tired of being forced to the margins of this country. With the courage to walk sixteen hundred kilometres through a Canadian winter to make the point that they will be Idle No More.

Francophones who live in Shediac, Sudbury, Saint Boniface, and all across this country who want their children to live and thrive in French, your determination inspires me; it must inspire the entire country.

Quebecers, from Gatineau to Gaspé, who want to re-engage with this country. With their country. Who have no time for the divisive issues of their parents' past, but want to work with Canadians who share their values to build a better country for all our kids.

I want to take a moment to speak directly to my fellow Quebecers.

Your engagement and your support in recent months has been deeply moving. I have learned so much from our conversations and our meetings. I take nothing for granted.

I understand that trust can only be earned. And my plan is to earn yours.

I feel confident about the future. I want to share with you why Quebecers have always been builders. From Champlain and Laurier to today, they have actively participated in shaping our country, together with so many other Canadians.

Our work is not complete. We face enormous challenges. Helping the middle class make ends meet. Reconciling economic growth and environmental stewardship. Playing a positive and meaningful role in the world. To rise above these challenges we must demonstrate our audacity and ambition, my friends. Audacity and ambition, always.

Let's be honest. We will not convince everyone. There will always be skeptics. People who say that our country is too big and too full of differences to be effectively managed, or for everyone to be represented. They are wrong, my friends.

I am not claiming that it will always be easy. That there will not be any obstacles along the way. That we will not have to make some compromises.

Canada is a grand, yet unfinished project. And it is up to us, together with all Canadians, to build the country that we want. The time has come for us to write a new chapter in the history of our country.

Let's leave to others the old quarrels and old debates that lead nowhere. Let's leave to others the ultra-partisan rhetoric and the old ways of doing politics. Let's leave the personal attacks to them.

Quebecers, let us be, together, once again, builders of Canada. So that our country can match the height of the dreams and ambitions that are shared across this country. So that we can leave our children a better world than the one that we inherited from our parents.

My friends, the Liberal Party will regain the confidence of Canadians when it proves that it is here to serve them. This is the task at hand. This is what will guide me as leader of the Liberal Party of Canada.

To the new generation of Canadians and to all the young people who are not engaged by politics, I have a very simple message for you: your country needs you. It needs your energy and your passion. It needs your idealism and your ideas.

The movement we are have been building over the past six months, it is yours. It belongs to you. It is the movement with which we will change politics. It is the movement that will allow us to reform our political institutions, to make reconciling the environment and our economy a real priority, and to play a positive and constructive role in the world.

My fellow Liberals, Canadians are looking to us. This campaign has been their campaign, more than just ours.

They want something better. They refuse to believe that better is not possible. They see the country their parents and grandparents worked so hard to build, and want to hand an even better country to their children.

Canadians share deep values that cannot be shaken, no

matter how hard the Conservative Party may try. Optimism. Openness. Compassion. Service to community. Generosity of spirit.

We want to believe that change can happen. We want leadership that will shape our best instincts into an even better country.

But Canadians will not suffer fools gladly. Canadians turned away from us because we turned away from them. Because Liberals became more focused on fighting with each other than fighting for Canadians.

Well, I don't care if you thought my father was great or arrogant. It doesn't matter to me if you were a Chrétien-Liberal, a Turner-Liberal, a Martin-Liberal, or any other kind of Liberal. The era of hyphenated Liberals ends right here, tonight.

From this day forward, we welcome all Liberals as Canadian Liberals. United in our dedication to serve and lead Canadians. Unity not just for unity's sake, but unity of purpose.

I say this to the millions of middle-class Canadians, and the millions more who work hard every day to join the middle class. Under my leadership, the purpose of the Liberal Party of Canada will be you. I promise that I will begin, spend, and end every day thinking about and working hard to solve your problems.

I know that you are optimistic about us, but cautiously so. You are, after all, Canadians. You know that hope is a fine thing, but that without an equal measure of hard work to back it up, it will be fleeting. So I know that you will judge us by the tenacity of our work ethic, the integrity of our efforts, and, come 2015, the clarity of our plan to make our country better. That is as it should be.

I know how lucky I have been in my life. Lucky, most of all, to have learned so much from so many Canadians. To learn that, above all else in this country, leadership means service.

I love this country, my friends, and I believe in it deeply. It deserves better leadership than it has now.

So let us be clear-eyed about what we have accomplished. We have worked hard and we have had a great campaign. We are united, hopeful, and resolute in our purpose.

But know this: we have won nothing more and nothing less than the opportunity to work even harder. Work even harder to prove ourselves worthy of leading this great country.

We should be deeply, deeply grateful for that opportunity. As your leader, I fully intend to make sure we make the most of it.

Change can happen. Canadians want leadership that will work with them to make it happen.

Be hopeful, my fellow Liberals. Work hard. Stay focused on Canadians. We can lead the change so many people want.

A better Canada is always possible. Together, we will build it.

Thank you.

Speech delivered at the 11th Annual Reviving the Islamic Spirit Convention

Toronto, December 22, 2012

As-salamu alaykum.

I am here today because I believe in freedom of expression.

I am here today because I believe in freedom of peaceful assembly.

I am here today because I believe in the Charter of Rights and Freedoms, which guarantees those sacred things to you, to me, and to all people with whom we share this land.

But mostly, I am here today because I believe in you. I believe in the contributions you have made to our country. And I know that together we will make even greater contributions in the future.

Let me begin with a story. A story from your history.

One that I hope will stay in your minds as you think about our common future.

Many generations ago, a young man was confronted by traditional religious elders. The kind of folks that today we might call fundamentalists or even extremists.

You see, a centuries-old conflict was raging. Prominent people on each side were convinced of their rightness, and loudly proclaimed that the other side was not only wrong, but wrong because of their religious beliefs, their culture, and their identity.

And as is far too often the case, these leaders reserved special scorn for those within their ranks who sought common ground with others. They understood the threat that moderation and compromise present to those who preach rigid doctrine.

This young man was struggling at the time. He was just starting out in the world. He was facing many of the same issues that, I suspect, you are facing today. How do I remain true to my values, to my culture, while I serve the interests of the society to which I belong?

He knew who he was, and what he believed. He was proud of his heritage, his culture, his religion. But he parted ways, decisively, with those within his community who would use these things to build walls.

But then, he was granted a remarkable opportunity, to address a distinguished audience of political, religious, and business leaders.

And so he challenged them to think beyond the narrow

confines of the present and to look toward the future. He said, "Providence has united together on this corner of earth populations of different origins and creeds. Is it not manifest that these populations must have together common and identical interests?"

That young man is a very important part of your history, as I said. But he would not go on to become an imam, a holy man, or a caliph.

He would, however, go on to become, among many more important things, my second-favourite prime minister.

The year was 1877. The place was Quebec City. And the brave young man's name was Wilfrid Laurier. He was thirty-five years old, with barely three years of service in Parliament to recommend him. And he had made a difficult choice.

Rather than fall in line with his elders and marshal his already prodigious talents in exclusive service of what he called his race, he chose an improbable new path. One that honoured what was good and noble about his own culture, yes. But one that used those very things to serve a higher purpose: to find common ground between people of differing beliefs.

Laurier saw something clearly, perhaps more clearly than any other Canadian: he saw that here, in this place, a new idea was taking shape. A new way of living together just might be possible.

He knew that his was a country founded and built by people who had warred against one another for centuries on their home continent: English versus French, Catholic versus Protestant. Early on, these murderous conflicts crossed the Atlantic Ocean with them.

But then a unique thing happened. Despite the fact that the English were victorious on the battlefield, the same measure of freedom was gained by each side.

In one of the most moving passages of that speech, speaking about the obelisk on the Plains of Abraham, Laurier said: "In what other country, under the sun, can you find a similar monument reared to the memory of the conquered as well as of the conqueror? In what other country, under the sun, will you find the names of the conquered and the conqueror equally honoured and occupying the same place in respect of the population? . . . Where is the Canadian who, comparing his country with even the freest countries, would not feel proud of the institutions which protect him?"

Now, the point of this story is not that remarkable moment in our history. The point is everything that has happened since.

This is our inheritance. One that has been renewed by successive generations to this very day.

That two peoples who had been enemies came together to build institutions—and a Constitution—that guaranteed freedom not only for one another but for all who would come after them.

They were joined in this great project over the years by people of every conceivable culture, religion, and ethnicity. Waves and waves of young men and women who chose to emphasize what was kind-hearted about their own traditions. Free people who chose to use the generosity of spirit that is the root of all faith to find common ground with those whose beliefs differed from their own.

As it is written in the Holy Qur'an: "The true servants of the Most Merciful are those who behave gently and with humility on earth, and whenever the foolish quarrel with them, they reply with [words of] peace" (Al-Furqan 25:63).

It has never been easy. This road has never been smooth or straight. Generations of Canadians had to overcome deep differences. They made a deliberate choice to turn their backs on rancour and conflict.

But today, because of them, we are all blessed to live in the most diverse country in the history of the world. One of the most peaceful and most prosperous.

One that has now moved beyond the goal of mere tolerance. Because saying "I tolerate you" is to grudgingly allow you to breathe the same air, to walk the same earth. And while there are many places in the world where tolerance is still just a far-off dream, in Canada we are beyond that. So let us not use the word *tolerance*. Let us speak instead of acceptance, understanding, respect, and friendship.

Here, we have come to a new realization, together: that a country can be great not in spite of its diversity, but because of its diversity.

This is our story now, yours and mine. The story of our country, Canada.

So as you reflect this weekend about the future, take

heart. Know that the struggles we are facing have been faced down before. Know that the conflicting feelings in our hearts have been felt before. Know that compromise and moderation are not the path of weakness but of courage and strength. That there is always a positive path in this country for all who seek common ground.

Most important, remember this: our inheritance must be constantly renewed by those who share Laurier's vision.

When people come together to create opportunities for one another, the dreams we hold in common will crowd out the fears that would divide us.

For it is not the political class but the middle class that unites this country. Open to all, our broad and diverse middle class is Canada's centre of gravity. Good people. People with common hopes and common challenges, coming together to find common ground.

There are already too many forces in the world that drive us into separate camps, that isolate us, and make us suspicious of one another.

Yesterday, protesters tried to prevent me from speaking at a school because of my stance defending gay marriage and women's rights. And as you know, some conservatives tried to stir up controversy about my appearance here today. They tried to appeal to people's fears and prejudices, the very things that this gathering was founded to overcome.

Now, I respect and defend their right to express their opinions. But I want you to know that I will always stand up to the politics of division and fear. It is short-sighted to pit groups of Canadians against one another. It may make some feel good for a little while, or even work politically in the short term.

But it is no way to build a country. Least of all this country. It is not who we are.

We are here today to do what we Canadians have been doing together for generations. We are honouring our diversity through friendship and understanding, so that we can build from it a common, positive future.

So I join you in your commitment to that more hopeful future. Let us pledge ourselves to building a country that brings people together; that finds the highest virtue in compromise, moderation, and common ground.

Nearly thirty years after that first speech, then in his third term as our prime minister, Laurier put it this way to an audience in Edmonton.

"We do not want or wish that any individual should forget the land of his origin. Let them look to the past, but let them still more look to the future. Let them look to the land of their ancestors, but let them look also to the land of their children. Let them become Canadians and give their heart, their soul, their energy, and all their power to Canada."

That was Laurier's wish for us. And it is mine for you. Be hopeful and positive, my friends. Your country needs you.

May peace, mercy, and blessings be upon you.

Acknowledgements

Tʜᴇʀᴇ ᴡᴇʀᴇ ᴍᴀɴʏ ᴘᴇᴏᴘʟᴇ ɪɴᴠᴏʟᴠᴇᴅ ɪɴ ᴛʜᴇ ᴄʀᴇᴀᴛɪᴏɴ of this book, and I'm grateful for their input and support.

My thanks to Jennifer Lambert, Iris Tupholme, Leo MacDonald, Michael Guy-Haddock, Sandra Leef, Cory Beatty, Rob Firing, Miranda Snyder, Noelle Zitzer, Neil Erickson, Alan Jones, Shaun Oakey, Sarah Wight, Anne Holloway, Michael Levine, Jonathan Kuy, John T reynolds, Reynolds, Caroline Jamet, Éric Fourlanty, Yves Bellefleur, Sandrine Donkers, Marie-Pierre Hamel, Brigitte Chabot, Joanna Gruda and Carla Menza, and everyone at Harper-Collins Canada and Les Éditions La Presse. They were all extremely patient about working around the craziness of my schedule and adjusting to the impossible pace I keep.

I'm grateful to my political team, who went above and beyond their usual responsibilities, especially Gerry Butts and Katie Telford but also Dan Gagnier, Cyrus Reporter, Alex Lanthier, Tommy Desfossés, Kate Purchase, Mylene Dupéré, and Kevin Bosch. They all helped in many, many different ways. Adam Scotti, photographer extraordinaire, shot many of the great images in this book and curated the rest.

Finally, and most importantly, thank you to Sophie, and to Xavier, Ella-Grace, and Hadrien, who put up with months of Dad working even harder than usual, often during family time that was already far too limited.

Any mistakes found in these pages are my own.

Photo Credits

———

All photos are provided courtesy of the author, except for those by Adam Scotti (3, 39, 44–71, 73–81) and the following:

17. The Canadian Press/Peter Bregg
18. Robert Cooper/Library and Archives Canada
25. Leslie Brock
32. Peter Bregg
33. Heidi Hollinger
34. Heidi Hollinger
35. Peter Bregg
36. Peter Bregg
42. Greg Kolz
43. Greg Kolz

Index